T0155510

SpringerBriefs in Applied Sciences and Technology

SpringerBriefs present concise summaries of cutting-edge research and practical applications across a wide spectrum of fields. Featuring compact volumes of 50–125 pages, the series covers a range of content from professional to academic.

Typical publications can be:

- A timely report of state-of-the art methods
- An introduction to or a manual for the application of mathematical or computer techniques
- A bridge between new research results, as published in journal articles
- A snapshot of a hot or emerging topic
- An in-depth case study
- A presentation of core concepts that students must understand in order to make independent contributions

SpringerBriefs are characterized by fast, global electronic dissemination, standard publishing contracts, standardized manuscript preparation and formatting guidelines, and expedited production schedules.

On the one hand, **SpringerBriefs in Applied Sciences and Technology** are devoted to the publication of fundamentals and applications within the different classical engineering disciplines as well as in interdisciplinary fields that recently emerged between these areas. On the other hand, as the boundary separating fundamental research and applied technology is more and more dissolving, this series is particularly open to trans-disciplinary topics between fundamental science and engineering.

Indexed by EI-Compendex, SCOPUS and Springerlink.

More information about this series at http://www.springer.com/series/8884

J. M. P. Q. Delgado · Ana Sofia Guimarães
António C. Azevedo · Romilde A. Oliveira
Fernando A. N. Silva · Carlos W. A. P. Sobrinho

Structural Performance of Masonry Elements

Mortar Coating Layers Influence

 Springer

J. M. P. Q. Delgado
CONSTRUCT-LFC, Faculty of Engineering
University of Porto
Porto, Portugal

Ana Sofia Guimarães
CONSTRUCT-LFC, Faculty of Engineering
University of Porto
Porto, Portugal

António C. Azevedo
CONSTRUCT-LFC, Faculty of Engineering
University of Porto
Porto, Portugal

Romilde A. Oliveira
Department of Civil Engineering
Catholic University of Pernambuco
Recife, Brazil

Fernando A. N. Silva
Department of Civil Engineering
Catholic University of Pernambuco
Recife, Brazil

Carlos W. A. P. Sobrinho
CONSTRUCT-LFC, Faculty of Engineering
University of Porto
Porto, Portugal

ISSN 2191-530X   ISSN 2191-5318   (electronic)
SpringerBriefs in Applied Sciences and Technology
ISBN 978-3-030-03269-2   ISBN 978-3-030-03270-8   (eBook)
https://doi.org/10.1007/978-3-030-03270-8

Library of Congress Control Number: 2018959265

This Springer imprint is published by the registered company Springer Nature Switzerland AG
The registered company address is: Gewerbestrasse 11, 6330 Cham, Switzerland

Preface

Over the last years, the occurrence of several accidents in the Metropolitan Region of Recife with masonry buildings constructed with non-structural blocks to carry loading beyond its own weight has drawn the attention of the regional and national technical community for the need to establish criteria of research, study, and rehabilitation, within acceptable levels of reliability. Masonry buildings constructed with such a technique are often referred to as resistant masonry buildings.

This book discusses masonry buildings constructed in the state of Pernambuco, Brazil. Topics such as the main features of this construction technique and the peculiarities that affect its structural behaviour are discussed. Technical information about accidents occurred in recent years is also discussed, along with the historical records of the events, followed by indications of the causes for the collapse.

Additionally, this work presents two experimental extensive campaigns with masonry elements in order to analyse the structural performance. First, an experimental study is carried out on running bond 195 red clay prisms, of two and three ceramic blocks, with and without cement mortar coating, subjected to axial compression in order to enhance the capacity of masonry. The prisms were subjected to compressive loading, and all of them had deformation control on each face with a deflectometer, in order to obtain information about the behaviour of the prisms. Secondly, an extensive characterisation of materials and components (more than 500 prisms made with ceramic blocks and concrete blocks) is used in non-structural masonry constructions in the region of Pernambuco, making it one of the most comprehensive research studies on this topic in Brazil. This study conducts an in-depth, numerical, and experimental analysis of the behaviour of the compressive strength of blocks, prisms, and mini-walls that are part of a non-load bearing system, which is often used in the region to carry loads above its own weight.

The main benefit of this book is the extensive experimental results obtained that allowed to identify the contribution of several mortar rendering layers to the load capacity of the tested specimens. The factors that influenced the load capacity of the tested specimens are also discussed.

Porto, Portugal   J. M. P. Q. Delgado
Porto, Portugal   Ana Sofia Guimarães
Porto, Portugal   António C. Azevedo
Recife, Brazil   Romilde A. Oliveira
Recife, Brazil   Fernando A. N. Silva
Porto, Portugal   Carlos W. A. P. Sobrinho

Contents

Chapter 1
Introduction

It is important in masonry design to determine the appropriate ultimate compressive strength of the masonry material. Masonry is a material built from units and mortar that induce an anisotropic behaviour for the composite. The lack of knowledge on the properties of the composite material imposes low assessments of the strength capacity of the masonry wall (Mohamad et al. 2012). Atkinson et al. (1985) state that the prediction of compressive strength and deformation of full scale masonry based on compressive tests of stack-bond masonry prism and the interpretation of the results of prism tests have a significant influence on the allowable stress and stiffness used in masonry design.

There have been numerous studies done on the behaviour of masonry prisms under axial compression. The effects of variables such as the height-to-thickness ratio of the prism, mortar type and grout strength, unit geometry, and various capping compounds have been the point of focus of many researchers. However, most of the research reports have been presented and published in various conferences around the world; however, some are unpublished (Aryana 2006). Moreover, some of the data that is the basis of the formula, the graphs, and the design tables presented in various parts of the Masonry Standards Joint Committee specification were the result of research done by the former Brick Institute of America, now the Brick Industry Association.

Structural masonry using hollow clay blocks has begun in Brazil in the mid 80s in residential constructions up to 6 storey high buildings. In the last two decades, clay structural masonry system has increased in some states of Brazil, mainly by the availability of blocks of high compressive strength produced in modular sizes. However, tests carried out in the last twenty years in Brazil (Cavalheiro and Gomes 2002) have shown that the average compressive strength of unreinforced clay walls is only 34% of the average compressive strength of the blocks, and the average compressive strength of two block prisms is 50% of the unit strength (Cavalheiro and Arantes 2004).

© The Author(s), under exclusive license to Springer Nature Switzerland AG 2019
J. M. P. Q. Delgado et al., *Structural Performance of Masonry Elements*, SpringerBriefs in Applied Sciences and Technology, https://doi.org/10.1007/978-3-030-03270-8_1

Some structural clay block producers may provide blocks of two or three resistances, with different prices, depending on the composition of clay mix, burning temperature, and even different cross sections varying coring patterns.

Despite the great interest, only a few studies have been carried out and published in Brazil on the influence of reinforced mortar coating on the compressive strength of clay brick masonry prisms.

1.1   Motivation

There have been numerous studies performed on the behaviour of masonry prisms under axial compression (Ewing and Kowalsky 2004; Kaushik et al. 2007a, b). The effects of variables such as the height-to-thickness ratio of the prism, mortar type and grout strength, unit geometry, and various capping compounds have been the point of focus of many researchers (Gumaste et al. 2007; Sumathi and Raja Mohan 2014).

The occurrence of several accidents in the Metropolitan Region of Recife—MR-R—with masonry buildings constructed with non-structural blocks to carry loading beyond its own weight has drawn the attention of the regional and national technical community for the need to establish criteria of research, study, and rehabilitation, within acceptable levels of reliability. Masonry buildings constructed with such technique is often referred to as resistant masonry buildings (see Fig. 1.1).

The resistant masonry is one constructive technique characterizes for the use of sealing units (ceramic or concrete) with structural purpose, supporting loads beyond its proper weight. The foundations generally are constructed in masonries with 9 cm or 19 cm of thickness, in continuity to the walls of the construction, usually seated on low shoe races of armed concrete with a transversal section in form of inverted T or on daily pay-moulded components of the foundation, seated on a layer of concrete regularization.

Diverse pathological manifestations have been observed, already having occurred, in some cases, collapses with fatal victims. It is important to register that the problem in quarrel does not consist in a local exclusiveness, and is of the knowledge of the authors the existence of accident with similar characteristics in Maceió and building pathology manifestations of the same nature in a situated building in Belo Horizonte (see Fig. 1.2).

Approximate numbers indicate that there are around 6000 residential buildings in the region made with this type of masonry buildings where close to 250,000 people live. Several pathological manifestations and collapses with human deaths have also been reported. The occurred accidents already in recent years lead to a probability of imperfection with superior values 1:500, when the socially acceptable one is off, in the maximum, 1:10,000. Twelve spontaneous landslides already had been registered, twelve buildings had been demolished and about 110 building if they find interdicted for not offering conditions of security for habitation (Oliveira and Sobrinho 2006; Oliveira et al. 2008).

Fig. 1.1 Examples of pathological manifestations observed, namely, a detail of masonry crushing between the foundation beam and ground floor slab

Rupture	Localization	Rupture cause
	Recife	Horizontal openings executed along the entire length of a central partition stop for the installation of conduits (1994)
	Jaboatão dos Guararapes	Loss of foundation block strength due to moisture expansion (1997)
	Olinda	Loss of resistance due to the degradation produced by the continuous action of sulfate ions on the concrete blocks (1999)
	Olinda	Rupture of the Foundation's ceramic blocks (1999) Seven fatalities occurred.
	Jaboatão dos Guararapes	The collapse of basements caused by the dismantling of running shoes due to the passage of sewage and rainwater (2011)
	Jaboatão dos Guararapes	There was a localized rupture of the basements in the region of the façade corresponding to the entrance of the building (2007)

Fig. 1.2 Examples of accidents occurred in the Metropolitan Region of Recife, Brazil

In this type of construction, generally leaked ceramic blocks seated with the perforations in the horizontal line or blocks of concrete are used, with low compressive strength (2.5 MPa). The frequency of these accidents and the brusque nature of the rupture, with gradual collapse, have generated fidget to the community technique and, mainly, to the inhabitants of these constructions, that today live in frightening for the uncertainty of the conditions of structural security of its residences.

The established framework, which constitutes a serious social problem that afflicts many developing countries, demands the carrying out of consistent scientific research that allows a deep understanding of the problem and allows the creation of technical retrofitting interventions to avoid new accidents.

This constructive practice had an important impetus from the beginning of the 60s and its success was due to the lower cost compared to the construction with conventional reinforced concrete structure and the velocity of execution in the region at the time (Oliveira et al. 2011).

On the other hand, the search for cost minimization, the lack of quality control of the components and the construction procedures, together with the lack of specific design codes has been causing a series of pathologies and accidents over the last years.

With regard to retrofitting strategies, there is scarce information in the literature on the subject. The only reference of which authors have knowledge is the research developed at the School of Engineering of São Carlos (EESC-USP), which investigated the contribution of the coating in the strength of masonry prisms built with non-structural blocks (Oliveira and Hanai 2002; Hanai and Oliveira 2006).

In local practice, what has been observed is the use of retrofitting solutions based on the empirical knowledge that needs more in-depth reflection on its effectiveness and applicability (Campos 2006).

In this context, the work presents results of research developed within the framework of the FINEP/HABITARE Project entitled Development of Models for Retrofitting Masonry Buildings Constructed with Non-Structural Bricks. The Project was conducted by the Catholic University of Pernambuco—UNICAP, as executing agency, by the Technological Institute of Pernambuco—ITEP, as proponent, by the Secretariat of Science, Technology and Environment—SECTMA, as an intervener and the University of Pernambuco—UPE and Federal University of Santa Catarina—UFSC—as co-executors.

The main objective of this work reports and discusses an extensive experimental campaign developed to understand the role of mortar coating on the compressive behaviour of unreinforced masonry elements, made with non-structural blocks. The experimental results allow evaluating the efficiency of several rehabilitation techniques.

References

S.A. Aryana, Statistical analysis of compressive strength of clay brick masonry prisms, M.Sc. Thesis, The University of Texas at Arlington (2006)

R.H. Atkinson, J.L. Noland, D.P. Abrams, S. McNary, A deformation failure theory for stack bond, brick masonry prisms in compression. Proceedings 3rd NAMC, Arlington, pp. 18-1 to 18-18 (1985)

M.D. Campos, Solutions considerations for recovery of constructed buildings with resilient masonry, M.Sc. Thesis in Civil Engineering, University Federal of Pernambuco, Recife, Brazil (2006)

O.P. Cavalheiro, C.A. Arantes, Influence of grout on hollow clay masonry compressive strength, in *13th International Brick and Block Masonry Conference*, Amsterdam, 4–7 July (2004)

O.P. Cavalheiro, N.S. Gomes, Structural masonry of hollow blocks: test results of elements and compressive strength reducers, in Proceedings of the 7th international seminar on structural masonry for developing countries, pp. 411–419, Belo Horizonte, Brazil (2002)

B.D. Ewing, M.J. Kowalsky, Compressive behavior of unconfined and confined clay brick masonry. J. Struct. Eng.—ASCE, 130(4), 650–661 (2004)

K.S. Gumaste, K.S. Nanjunda Rao, B.V. Venkatarama Reddy, K.S. Jagadish, Strength and elasticity of brick masonry prisms and wallettes under compression. Mater. Struct. 40(2), 241–253 (2007)

J.B. Hanai, F.L. Oliveira, Collapsing of ceramic block buildings. Téchne 115, 58–63 (2006)

H.B. Kaushik, D.C. Rai, S.K. Jain, Stress-strain characteristics of clay brick masonry under uniaxial compression. J. Mat. Civil Eng.—ASCE, 19(9), 728–739 (2007a)

H.B. Kaushik, D.C. Rai, S.K. Jain, Uniaxial compressive stress–strain model for clay brick masonry. Curr. Sci. 92(4), 497–501 (2007b)

G. Mohamad, P.B. Lourenço, H.R. Roman, C.S. Barbosa, E. Rizzatti, Stress-strain behaviour of concrete block masonry prisms under compression. 15th international brick and block masonry conference, Florianópolis, Brazil (2012)

F.L. Oliveira, J.B. Hanai, Behaviour analysis of masonry walls constructed with ceramic sealing blocks, in *Proceedings of the International Seminar on Structural Masonry for Developing Countries*, Belo Horizonte, Brazil (2002)

R.A. Oliveira, F.A.N. Silva, C.W. Sobrinho, Buildings Constructed with resistant Masonry in Pernambuco: current situation and future prospects. In: B.S. Monteiro, J.A.P. Vitório (Organizers). SINAENCO-PE and knowledge production: collection of technical articles, Recife, Brazil (2008)

R.A. Oliveira, C.W. Sobrinho, Accidents with buildings constructed with resistant masonry in the Metropolitan Region of Recife. João Pessoa: DAMSTRUC, Recife, Brazil (2006)

R.A. Oliveira, F.A.N. Silva, C.W. Sobrinho, Resistant Masonry: an experimental and numerical investigation of its compressive behaviour, Recife: FASA, Brazil (2011)

A. Sumathi, K. Saravana Raja Mohan, Study on the effect of compressive strength of brick masonry with admixed mortar. Int. J. ChemTech Res. 6(7), 3437–3450 (2014)

Chapter 2
Physical and Hygrothermal Material Properties

2.1 Ceramic Clay Bricks

The geometrical characteristics of the clay blocks, i.e., their shape and manufacturing dimensions, must meet the tolerances provided:

– Face measures—Effective dimensions;
– The thickness of the septa and external walls of the blocks;
– Deviation from the square (D);
– Face plane (F);
– Gross area (Ab).

The apparatus required to carry out the measurements consisted of pachymeter with a minimum sensitivity of 0.05 mm, a metal ruler with a minimum sensitivity of 0.5 mm, a metal bracket of 90 + 0.5° and a balance with a resolution of 10 g, all of them were properly calibrated. The measurements of the block faces, the values of width (W), height (H) and length (L) were determined as shown in Fig. 2.1.

The septa measurement had the following procedure and the measurements were made in the central region of the clay blocks, using at least four measurements, searching for the narrowest septa, as we shown in Fig. 2.2a. The flatness of the faces was determined by the arrow formed in the diagonal of one of the facing faces of the block, according to Fig. 2.2b.

Finally, the value of the gross area of each clay block was determined as shown in Table 2.1, together with all the data of the geometric characterization.

© The Author(s), under exclusive license to Springer Nature Switzerland AG 2019
J. M. P. Q. Delgado et al., *Structural Performance of Masonry Elements*, SpringerBriefs in Applied Sciences and Technology, https://doi.org/10.1007/978-3-030-03270-8_2

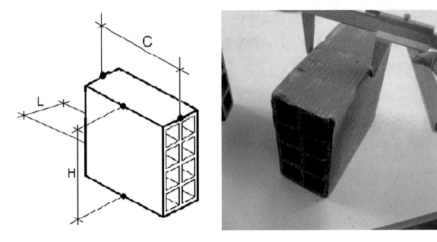

Fig. 2.1 Septa measurement

(a)   **(b)**

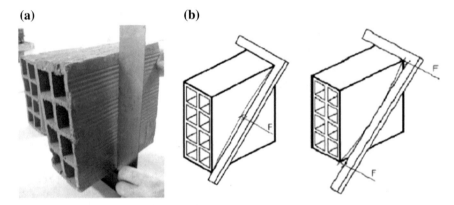

Fig. 2.2 **a** Measurement of the deviations and **b** planar measuring (*source* NBR15270-3)

2.1.1   Physical Characteristics

The physical characteristics of the ceramic blocks analysed were:

- Dry mass (m_{dry});
- Water absorption index (AA);
- Initial absorption index (AAI).

The equipment used for such determinations were: balance with a resolution of up to 5 g, oven with adjustable temperature and a tank for submersion of the samples. The number of samples established by the standard to determine these characteristics is equal to six specimens.

Table 2.1 Geometric characterization of the ceramic blocks tested

Specimens		Face measures—effective dimensions (mm)			Thickness of the septa and external walls (mm)		Square deviation (mm)	Face plane (mm)	Gross area (cm²)
		Width (W)	Height (H)	Length (L)	Septa	External walls	D	F	Ab
1		93	193	190	7	8	1	1	177
2		91	191	191	7	8	1	1	173
3		93	192	190	8	8	1	1	176
4		90	192	190	8	8	1	1	171
5		91	190	188	8	8	1	1	172
6		91	190	191	7	8	1	1	175
7		91	191	189	8	9	1	1	173
8		90	190	191	8	8	1	1	172
9		92	188	190	7	8	1	1	174
10		92	192	191	8	9	1	1	176
11		92	191	192	8	8	1	1	175
12		89	190	188	7	8	1	1	167
13		91	189	193	7	8	1	1	175
Average		91	191	190	8	8	1	1	174
Tolerance NBR 15270-1 (mm)	Individual ±	5	5	5	Minimum 6	Minimum 7	Maximum 3	Maximum 3	x
	Average ±	3	3	3	0	0	0	0	x
Non-conforming units		0	0	0	0	0	0	0	x

Reference dimensions of the clay blocks: 90 × 190 × 190 mm

(a)   **(b)**

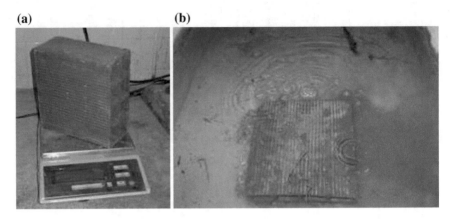

Fig. 2.3 **a** Wet weight and **b** submerged specimen

To determine the dry mass (m_{dry}) the samples of the ceramic blocks were first cleaned for dust removal and other loose particles adhered to the block and identified, after which they were submitted to a temperature range of $(105 + 5)$ °C up to the stabilization of the individual mass, when after two consecutive weighing, with intervals of 1 h, did not differ by more than 0.25% of the weight, as shown in Fig. 2.3a.

After the determination of the dry mass, the blocks were completely submerged in a tank with water at room temperature for a period of 24 h and then removed and placed in a balance to determine the wet mass (m_{wet}) after removal of excess water with a cloth (see Fig. 2.3b).

The water absorption index (AA) of each specimen was calculated using the following expression:

$$AA(\%) = \left[\left(m_{wet} - m_{dry}\right) + m_{dry}\right] \times 100 \qquad (2.1)$$

The determination of the initial absorption index (IRA) required, in addition to the equipment described above, a chronometer with a sensitivity of 1 s, a bubble level ruler, a water reservoir that allows the maintenance of a $(3 + 0.2)$ mm as shown schematically in Fig. 2.4.

The samples were first subjected to heating in an oven for 24 h and after their removal 2 h were allowed for cooling in the open air until the ambient temperature. The geometric characteristics of the blocks were determined to obtain the contact area with the water slide. The blocks were then positioned on the supports so that the contact face of the block remained in contact with the water slide at a height of 3 mm for a time of 1 min, then the block was removed and removed the excess water with the aid of a damp cloth to proceed with the weighing of the block.

The IRA is the water absorption index (suction) of the tested face of the blocks, expressed in (g/193.55 cm^2)/min and is calculated according to the expression:

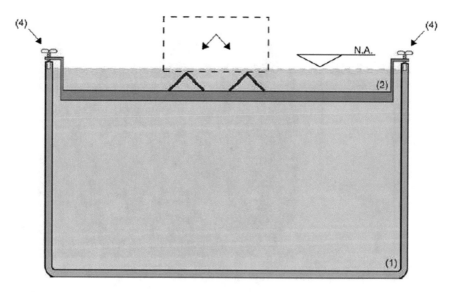

Fig. 2.4 Sketch of IRA determination (*source* NBR15270-3)

Table 2.2 Properties of the clay brick blocks used

Samples	IRA[a]	Dry mass (g)	Compressive strength[b] (MPa)
1	6.9	2406	2.17
2	12.6	2320	2.10
3	9.2	2289	2.16
4	8.6	2338	2.12
5	17.4	2307	2.02
6	3.9	2403	2.04
Average	9.77	2331.4	2.05
Standard deviation	±4.70	±43.86	±0.40
COV[c]			19%

$$IRA = 19.55 \times (\Delta P + area) \qquad (2.2)$$

where ΔP is the wet mass change after 1 min (dried at room temperature). The properties of the clay bricks used in prism construction are presented in Table 2.2. All the samples used in testing had a net area that exceeded 75% of their gross area.

(a)   **(b)**

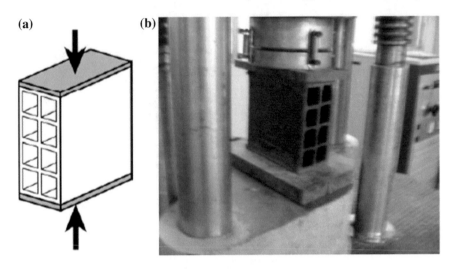

Fig. 2.5 **a** Load application axis and **b** Specimens before the experimental tests

2.1.2   Mechanical Characteristics

The mechanical characteristics of the ceramic clay brick blocks were evaluated by individual compressive strength. For the accomplishment of this test, a total of 13 specimens were properly prepared. The regularization of the two faces destined to the settlement perpendicular to the block length was done with cement and a maximum thickness of 3 mm in order to uniformity the block surfaces.

After hardening the capping layers, the specimens were completely submerged in a tank with water for a period not lesser than 6 h, as established in the Brazilian standard NBR 15270-3 (2005). The machine used to carry out the tests was the Universal Testing Machine a facility of the Laboratory of Construction Materials of Catholic University of Pernambuco, Brazil. The specimens were tested in saturated conditions and placed in the press so that their center of gravity coincides with the load axis of the press plates, as illustrates in Fig. 2.5. The results obtained are presented in Table 2.1.

2.2   Sand

2.2.1   Fine Aggregate Analyse

The characterization of the fine aggregates (sand), used in the preparation of mortars, and took into account the following tests and their respective Brazilian standards:

– Granulometry of the fine aggregate—NBR NM 248 (2003)
– Dry aggregate specific gravity (Flask of Chapman)—NBR 52 (2009)

Fig. 2.6 Sieves used

- Aggregate specific gravity in (SSS)[3] conditions—NBR NM 52 (2009)
- Bulk density—NBR 52 (2009)
- Determination of the fine materials content—NBR NM 46 (2003)
- Fineness modulus—NBR NM 248 (2003)
- Maximum diameter—NBR NM 248 (2003)
- Determination of the unit mass—NBR NM 45 (2006)
- Water absorption—NBR NM 30 (2001).

The sand used was acquired in the Metropolitan Region of Recife and all the tests were carried out in the Laboratory of Construction Materials of Catholic University of Pernambuco—TECOMAT, Brazil.

2.2.1.1 Determination of the Granulometric Composition

The determination of the granulometric composition aims to classify the aggregate as a function of the size of the grains. The test method is described in NBR NM 248 (2003) and to determine if it is necessary to collect two samples of the aggregate to be analysed which must then be washed and preheated to a temperature of (105 + 5) °C. Its classification occurs through a set of sieves with openings standardized by ABNT. The assembly is organized in a decreasing manner so that the sieves of larger apertures overlap the sieves of smaller apertures; the material is then sieved so that each fraction is retained in the sieves, and then separated and weighed. Figure 2.6 illustrates the procedure.

2.2.1.2 Determination of the Specific Mass

In order to determine the specific mass of the fine aggregate, the procedure described in the Brazilian standard NBR NM 52 (2009), using the Chapman vial (see Fig. 2.7),

Fig. 2.7 Chapman's vessel

a standardized vial, that indicates the displacement of the water column volume after the materials insertion, was used. The material specific mass is the ratio of its mass and the volume occupied by it, excluding voids between the grains.

The sand was oven dried at a temperature of about 105 °C and a sample of 500 g was withdrawn. This sample was then inserted into the Chapman vessel, which was already filled with water until a volume of 200 cm^3; as the material was being placed, the water column moved so that at the end, the variation of this displacement represented the volume of sand inserted. The equation used is shown below:

$$\gamma = m \div (L - 200) \tag{2.3}$$

where m is the dry mass of the material and L is the final volume of the water column, the result is expressed in g/cm^3.

2.2.1.3 Determination of the Fine Materials Content

Fine materials present in the aggregate, i.e., those passing through the 75 μm aperture sieve, also called powder material, should be analysed by the method described in NBR NM 46 (2003). Its quantity, when higher than the one foreseen in the standard

NBR 7211 (2009) that is of 5%, can harm the mixture, either of concrete or mortar because the very fine grains make difficult the adhesion of the paste of cement to the aggregate.

The sample was oven-dried at 110 °C until mass constancy and then a 100 g sample was withdrawn. This quantity was placed on the sieves with opening 1.2 mm and 75 μm and subjected to the washing successive times so that the fine material adhered to the aggregate was eliminated with the water. The process was completed when the water passed through was completely clean. The material was again placed in an oven to evaporate the water and obtain the final mass. The result was calculated according to the following expression:

$$\text{Powder material content} = \left[\left(M_i - M_f\right) \div M_i\right] \times 100 \tag{2.4}$$

where M_i is the initial mass and M_f is the final mass.

2.2.1.4 Clay Content in Clods

Clad clays and friable materials are materials that are susceptible to wear when subjected to minor stresses that may alter the quality of an aggregate for contamination with poorly resistant grains and which impair both the strength and the appearance of concrete and mortars. Its content is calculated according to the recommendations code NBR 7218 (2010), separating a portion of the kid's aggregate that passes in the sieve with opening 4.8 mm and is retained in the sieve with opening 1.2 mm, identifying the clods and friable grains and proceeding with discharging of these grains and subjecting them to a new sifting process. The calculation is based on the following equation:

$$\text{depleted material} = \left(m_i - m_f\right)/m_i \tag{2.5}$$

where m_i is the initial mass and m_f is the final mass after sieving.

2.2.1.5 Unit Mass Test

The unit mass in the loose state of the fine aggregate is determined by NBR NM 45 (2006). In this method a cylindrical vessel with a known volume of 20 l, 16 mm diameter metal rod, a concrete shovel, metal ruler, a balance with a resolution of 50 g and an amount of dry sand sufficient to occupy the volume of the vessel.

In this test, method C of the standard was used. The dry sand is added without compaction until the entire volume of the vessel is occupied, the metal ruler is used to remove the excess sand on the vessel and the sand set plus vessel is weighed, the result is calculated according to the expression:

$$\rho_{ap} = (m_{ar} - m_r) \div V \tag{2.6}$$

where ρ_{ap} is the unit mass of the aggregate (kg/m^3), m_{ar} is the mass of sample plus container (kg), m_r is the empty container mass (kg) and V is the container volume (m^3).

2.2.1.6 Swelling of the Fine Aggregate

The swelling is a phenomenon that concerns the small aggregate and can be described as the variation of the apparent volume that affects the unit mass of the material when it is submitted to the variation of the moisture content. In other words, the same amount of sand may occupy larger or smaller volumes without compaction when its moisture content varies from dry to wet sand. When performing a volume trace, it is very important to correct the volume of the sand in the container as well as its moisture content, due to the great capacity of water retention by the small aggregates.

The test, according to the standard NBR 6467 (2009), is to perform several measurements of the unit mass in various sand moisture conditions, namely: 0, 0.5, 1, 2, 3, 4, 5, 7, 9 and 12% and the coefficient of swelling is then calculated by the expression:

$$CI = (V_h \div V_o) = \left[(\gamma_s \div \gamma_h) \times ((100 + h) \div 100) \right] \qquad (2.7)$$

where V_h is the volume of the aggregate with h% humidity (dm^3), V_0 is the volume of dry aggregate in greenhouse (dm^3), V_h/V_0 is the swelling coefficient, γ_s is the unit mass of the dry aggregate (kg/dm^3), γ_h is the aggregate mass with h% of moisture content (kg/dm^3) and h is the aggregate moisture content (%).

2.2.1.7 Water Absorption

The test aims at determining the level of water absorption in small aggregates, for a specific use. It's extremely relevant to specifically get from the aggregate its degree of absorption which will be used in a concrete trace, for instance. In this case, when aggregate already shows some moisture percentage the amount of water usually added to the paste should be decreased. Therefore, an eventual segregation will be prevented, where larger particles will separate from smaller ones, thus causing the emergence of layers in the concrete and consequently reducing seriously its resistance. The test method, described in NBR NM 30 (2001), consists of quartering 1 kg of an aggregate sample being analyzed which must be washed and placed in advance in the greenhouse at a temperature of (105 ± 5) °C for 6 h.

2.2.1.8 Sand Analyse Results

The sand used in the preparation of the mortars for laying and coating the tested models is usually found in the Metropolitan Region of Recife (MRR) and all the lot used in the development of the research was acquired from the same supplier.

The results of the granulometric composition obtained are presented in Tables 2.3 and 2.4, for the fine aggregate, according to the Brazilian standard NBR NM 248 (2003).

Table 2.5 presents the granulometric results of coarse aggregate. Table 2.6 presents a summary of the characterization results obtained with the fine aggregates used in the research.

2.3 Mortar

The mortars used in this work were subject to a detailed analysis process in which their properties were investigated in the fresh state as in the hardened state. Table 2.7 shows the grout mixtures (mix ratio) studied as well as their applications.

The fresh mortars were characterized by consistency and density test. The mortar consistency was evaluated through the procedures described in the Brazilian standard NBR 7215 (1994), allowing to verify the plasticity degree with a Flow Table. The samples were subjected to successive fall from a pre-established height so that the more plastic the mortar, the greater its final diameter.

A mortar may be considered dry when the consistency index (flow table) is less than 250 mm. Mortars with a consistency index between 260 and 300 mm (ex. plaster mortar) are considered plastic. Finally, mortars with a consistency index of more than 360 mm are considered to be fluid. The results are presented in Table 2.7.

The mortars in the hardened state were characterized by the following tests:

– Axial compressive strength (rupture test), NBR 5738 (2004);
– Diametrical compressive tensile strength, NBR 7222 (1994).

In order to measure the axial compressive strength, six cylindrical specimens were prepared for each mortar type, with a diameter of 5 cm and a height of 10 cm. The specimens were cured for a period of 28 days, before carrying out the resistance tests, as described in Table 2.7.

In order to perform diametrical compressive tensile strength tests, the procedures described in the Brazilian standard NBR 7222 (1994) were used, which establishes the preparation of six specimens, in the same way in which they were prepared for the axial compression resistance tests. The specimens were positioned in order that the contact between the plates of the test machine and the specimens gives only two generators diametrically opposite to the specimen. The results obtained are presented in Table 2.7.

The mortars used both in the laying of the blocks and in the coating were defined from cement, lime and sand mixtures in proportions of 1:2:9, 1:1:6 and 1:0.5:4.5 by volume.

Table 2.3 Granulometric results of the fine aggregates (NBR NM 248 2003)

Opening sieves (mm)	Maas retained (%)		Variation of the % retained ≤ 4% (%)	Average of retained mass (%)	Cumulated of retained mass (%)	Ratios over % retained			
						Lower limits		Higher limits	
	Test no 1 (%)	Test no 2 (%)				Used zone	Optimal zone	Used zone	Optimal zone
9.5	0.0	0.0	0.0	0.0	0.0	0	0	0	0
6.3	0.0	0.0	0.0	0.0	0.0	0	0	0	7
4.75	0.6	0.3	0.2	0.5	0.5	0	0	5	10
2.36	4.1	3.3	0.8	3.7	4.1	0	10	20	25
1.18	10.1	8.9	1.2	9.5	13.6	5	20	30	50
0.6	21.4	21.4	0.0	21.4	35.1	15	35	55	70
0.3	33.4	32.5	0.9	33.0	68.0	50	65	85	95
0.15	21.2	23.1	1.9	22.1	90.2	85	90	95	100
Bottom =	9.2	10.5	1.2	Fine module =	2.11	Maximum characteristic size (mm) =			2.4

Table 2.4 Summary of the characterization results obtained with the fine aggregates

Specific mass (g/cm^3)	Clay clods (%)	Fine material (%)	Unitary dry mass (kg/m^3)	Swelling	
				Swelling (%)	Critical humidity (%)
2.62	0.0	4.4	1450	1.23	3.20
NBR 9776 (1987)	NBR 7218 (2010)	NBR NM 46 (2003)	NBR NM 45 (2006)	NBR 6467 (2009)	

Table 2.5 Granulometry results of coarse aggregate—NBR NM 248 (2003)

Opening of the sieves (mm)	Mass retained (g)		Average of retained mass (g)	Average of retained mass (%)	Cumulated of retained mass (%)
	Test n° 1	Test n° 2			
25	0	0	0	0	0
19	0	0	0	0	0
12.5	0	0	0	0	0
9.5	0	0	0	0	0
6.3	0	0	0	0	0
4.75	1	1.08	1.04	0.1	0
2.36	2.1	2.16	2.13	0.3	0
1.18	9.7	9.62	9.66	1.3	2
0.6	274.4	274.33	274.37	37.2	39
0.3	249	249.09	249.05	33.8	73
0.15	112.6	212.55	112.58	15.3	88
Bottom	88.8	88.87	88.84		
Total	737.6	737.7	737.7		

Table 2.6 Experimental results of the aggregate characterization

	Dry aggregate specific gravity (g/cm^3)	Aggregate specific gravity in (SSS)^3 conditions (g/cm^3)	Bulk density (g/cm^3)	Powder material (%)	Fineness modulus	Maximum diameter (mm)	Water absorption (%)
Fine aggregate	2.53	2.55	2.60	0.5	3.35	4.75	1.2
Coarse aggregate	2.42	2.44	2.48	2.7	2.02	2.02	1.0
	NBR NM 52 (2009)	NBR NM 52 (2009)	NBR NM 52 (2009)	NBR NM 46 (2003)	NBR NM 248 (2003)	NBR NM 248 (2003)	NBR NM 30 (2001)

Table 2.7 Consistency index, axial compression strength and tensile strength by diametrical compression of the mortars tested

Mortar application	Grout mixtures (mix ratio)	Consistency index (mm)	Axial compressive strength (kN)	Tensile strength (MPa)
Settlement	1:1:6 (cement:lime:sand)	266	6.5 ± 1.1	0.9 ± 0.2
Roughcast	1:3 (cement:sand)	305	30.4 ± 0.6	2.8 ± 0.9
Coating	1:2:9 (cement:lime:sand)	302	2.8 ± 0.4	0.8 ± 0.3
Coating	1:1:6 (cement:lime:sand)	306	6.5 ± 1.1	0.9 ± 0.2
Coating	1:0.5:4.5 (cement:lime:sand)	296	5.4 ± 0.5	0.9 ± 0.8

References

NBR 15270-3, *Structural and Non-structural Ceramic Blocks—Test Methods*, Rio de Janeiro, Brazil (2005)

NBR NM 52, *Fine Aggregate—Determination of the Bulk Specific Gravity and Apparent Specific Gravity*, Rio de Janeiro, Brazil (2009)

NBR NM 30, *Fine Aggregate—Test Method for Water Absorption*, Rio de Janeiro, Brazil (2001)

NBR NM 46, *Aggregates—Determination of Material Finer Than 75 μm Sieve by Washing*, Rio de Janeiro, Brazil (2003)

NBR NM 45, *Aggregates—Determination of the Unit Weight and Air-Void Contents*, Rio de Janeiro, Brazil (2006)

NBR 7211, *Aggregate for Concrete—Specification*, Rio de Janeiro, Brazil (2009)

NBR 7218, *Aggregates—Determination of Clay Lumps and Fiable Materials Content—Method of Test*, Rio de Janeiro, Brazil (2010)

NBR 6467, *Aggregates—Determination of Swelling in Fine Aggregates—Method of Test*, Rio de Janeiro, Brazil (2009)

NBR NM 248, *Aggregates—Sieve Analysis of Fine and Coarse Aggregates*, Rio de Janeiro, Brazil (2003)

NBR 9776, *Aggregate—Determination of Fine Aggregate Specific Gravity by Chapman Vessel—Method of Test*, Rio de Janeiro, Brazil (1987)

NBR 7222, *Mortar and Concrete—Determination of the Tension Strength by Diametrical Compression of Cylindrical Test Specimens*, Rio de Janeiro, Brazil (1994)

NBR 5738, *Concrete—Procedure for Moulding and Curing Concrete Test Specimens*, Rio de Janeiro, Brazil (2004)

NBR 7215, *Portland Cement—Determination of Compressive Strength*, Rio de Janeiro, Brazil 35 (1994)

NBR15270-1, *Ceramic Components Part 1:Hollow Ceramic Blocks for Non-Load BearingMasonry— Terminology and Requirements*, Rio de Janeiro, Brazil (2005)

Chapter 3
Influence of Reinforced Mortar Coatings on the Compressive Strength of Masonry Prisms

3.1 Materials

This work describes an experimental study carried out on running bond 195 red clay prisms, of two and three ceramic blocks, with and without cement mortar coating, subjected to axial compression in order to enhance the capacity of masonry. The prisms were subjected to compressive loading and all of them had deformation control on each face with a deflectometer, in order to obtain information about the behaviour of the prisms.

Experimental tests with red clay prisms were performed in the Laboratory of Construction Materials of Catholic University of Pernambuco, Brazil. The red clay brick blocks used had dimensions of $91 \times 191 \times 190$ mm^3 (with a tolerance of ± 3 mm) and an average density of 2620 kg/m^3. Overall, a total of 195 prisms were built and tested (see Table 3.1). All the applicable ASTM standards or Brazilian standards (NBR) were followed in the building, curing, capping, and testing of the prisms and the components.

In order to perform the mechanical deflectometer readings, 400 L-shaped metallic plates, with 6 cm high and 2 cm wide, were fabricated (see Fig. 3.1), which were used as a support base for the deflectometers. The metallic plates were fixed to the prisms, previously, in the middle third of their length, through bonding, at a distance sufficient to allow the free flow of the deflectometers, as seen in Fig. 3.1a.

The hydraulic jacks had 200 mm piston stroke and 50-ton load capacity. This allowed only one jack to be sufficient to apply the load required for rupture of the prisms. The load drive machine, controlled by software which allows a perfect control of both displacement increment and load increase, has a servo-hydraulic working system and it is connected to the linear displacement sensors (LVDT). The displacement control of the hydraulic jacks makes it possible not only to follow the post-cracking and post-rupture, but also the shape of the rehab curve of the samples in front of the maintenance or increase of displacement.

© The Author(s), under exclusive license to Springer Nature Switzerland AG 2019
J. M. P. Q. Delgado et al., *Structural Performance of Masonry Elements*, SpringerBriefs in Applied Sciences and Technology, https://doi.org/10.1007/978-3-030-03270-8_3

Table 3.1 Description of the tested samples

Ref.	Samples tested	Dimension (cm × cm)	Area (cm^2)
Prims with 2 clay blocks			
2P-1	Uncoated prims	9 × 19	171
2P-3	Prisms with a coating of 3 cm of mix ratio 1:1:6 (cement:lime:sand)	15 × 19	285
2P-5	Prisms with a coating of 3 cm and mix ratio 1:2:9 (cement:lime:sand)	15 × 19	285
2P-7	Prisms with a coating of 3 cm and mix ratio 1:1:6 (cement:lime:sand) reinforced with a POP mesh of 10 × 10 cm and 4.2 mm of diameter	15 × 19	285
Prims with 3 clay blocks			
3P-1	Uncoated prisms	9 × 19	171
3P-2	Prisms with a mix ratio of 1:3 (cement:sand)	10 × 19	190
3P-3	Prisms with a coating of 3 cm of mix ratio 1:1:6 (cement:lime:sand)	15 × 19	285
3P-4	Prisms with a coating of 1.5 cm and mix ratio 1:2:9 (cement:lime:sand)	12 × 19	228
3P-5	Prisms with a coating of 3 cm and mix ratio 1:2:9 (cement:lime:sand)	15 × 19	285
3P-6	Prisms with a coating of 3 cm of mix ratio 1:0.5:4.5 (cement:lime:sand)	15 × 19	285
3P-7	Prisms with a coating of 3 cm and mix ratio 1:1:6 (cement:lime:sand) reinforced with a POP mesh of 10 × 10 cm and 4.2 mm of diameter	21 × 19	399
3P-8	Prisms with a coating of 1.5 cm and mix ratio 1:2:9 (cement:lime:sand) reinforced with a POP mesh of 10 × 10 cm and 4.2 mm of diameter	18 × 19	342
3P-9	Prisms with a coating of 3 cm and mix ratio 1:2:9 (cement:lime:sand) reinforced with a POP mesh of 10 × 10 cm and 4.2 mm of diameter	21 × 19	399

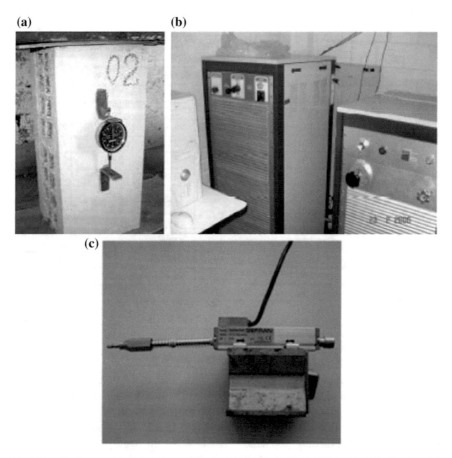

Fig. 3.1 **a** Deflectometer; **b** Monitors drive machine and **c** LVDT (Linear Variable Differential Transformer)

In this research, we promote two distinct, but simultaneous, ways of measuring the displacements. The first one used two displacement sensors, or LVDT's (Linear Variable Differential Transformer), a particular type of variable reluctance sensor (see Fig. 3.1c). The second way of displacement measurement came from the need to compare the measurements obtained with LVDT's, which take into account the entire length of the prism along with the wood deformation, with measurements of section parts of the prism length. Several procedures were tested and it was decided to measure the middle third. In this way, two pre-bonded metal plates were placed on each side of the prism to serve as support for the mechanical extensometers.

Table 3.2 Prisms test results

Sample	Thick (cm)	Width (cm)	Height (cm)	Prisms—Compressive resistance (kN)		
				Average	St. Dev.	Cov (%)
2P-1	9	19	39	9.72	3.38	35
2P-3	15	19	39	45.47	12.21	27
2P-5	15	19	39	50.43	13.61	27
2P-7	15	19	39	120.73	16.89	14
3P-1	9	19	59	9.49	2.08	22
3P-2	10	19	59	18.05	5.92	33
3P-3	15	19	59	52.71	9.07	17
3P-4	12	19	59	39.58	9.22	24
3P-5	15	19	59	45.03	10.38	23
3P-6	15	19	59	59.02	8.81	15
3P-7	21	19	59	109.17	11.23	10
3P-8	18	19	59	94.47	12.17	13
3P-9	21	19	59	100.25	10.54	11

3.2 Results

Table 3.2 shows that the average compressive resistance of the uncoated prisms with two (2P-1) or three blocks (3P-1) was not very different. In fact, for the two-block prism, the mean load was 9.72 kN whereas for the three-block prisms this load was 9.49 kN, representing a difference of 2.4%. However, it is important to note that the coefficients of variation observed were 35 and 22%, for uncoated prisms with two and three blocks respectively, indicate that the observed values should be examined with caution. In addition, it should be noted that these prisms were tested on a hydraulic press machine with a load capacity of 500 kN and the average compressive resistance values obtained were below the accuracy of the press machine, which is on the order of 10% of its capacity. In spite of these aspects, it is possible to conclude that no significant difference was observed between prisms of two and three blocks.

Coated prisms made with weak and medium mortar, with 2 and 3 blocks, according to Table 3.2, had a significant increase of load capacity, reaching up to ~420% for 2 block prisms and ~450% for 3 block prisms.

For the 2 block prisms with a coating, it was observed that the prisms with medium mortar presented lower compressive load capacity than the prisms made with weak mortar, although the difference did not exceed 10%. The coefficient of variation observed for these two situations was of the same order of magnitude, approximately 27%, is this value considered statistically high, possibly explaining the unexpected result.

Table 3.3 Wallettes test results

Sample	Thick (cm)	Width (cm)	Height (cm)	Wallets—Compressive resistance (kN)		
				Average	St. Dev.	Cov (%)
W-1	9	59	119	56.3	8.7	15.4
W-2	10	59	119	84.9	16.3	19.2
W-3	15	59	119	168.3	33.3	19.8
W-4	12	59	119	130.4	24.4	18.7
W-5	15	59	119	156.5	16.1	10.3
W-6	15	59	119	262.2	42.7	16.3
W-7	21	59	119	417.1	63.0	15.1
W-8	18	59	119	321.0	47.7	14.9
W-9	21	59	119	367.0	49.3	13.4

The observed inconsistency in the mean burst load of the 2 block prisms was not observed in the 3 block prisms. This can be justified by the fact that the coefficient of variation for the 3 block prisms was lower than for 2 block prisms.

The grout mixtures and the thickness of the mortar in the 3 block prisms had a significant influence on the compressive resistance of these elements. It is possible to observe in prisms with a single coating mortar (1:1:6), that increasing the thickness from 1.5 to 3 cm the load capacity increased by approximately 14%. While for prisms with the same thickness of 3 cm and different mix ratio it was observed and an increase of 31% between mix ratios of 1:1:6 and 1:0.5:4.5.

Reinforced prisms with 3 cm of thickness reinforced mortar and connectors showed a significant increase in the load capacity when compared to the prisms without framework and connectors. For the prisms with 2 blocks and a medium mortar with 3 cm of thickness the observed increase was 165%. For the 3 block prisms with weak mortar, 1.5 cm or 3 cm thickness, this increase was 139 and 123%, respectively. For the 3 block prisms with medium mortar, 3 cm thickness, the increase was 107%.

The wallets tested present a height of 119, 59 cm of width and thickness that depends of the coating used. Tables 3.3 and 3.4 also present a comparison between the average compressive resistance of prisms with 2 and 3 blocks and ceramic wallettes. It is worth noting that the load application area of the prism is 19×9 cm^2 and for the wallettes is 59×9 cm^2, which implies a wall/prism area ratio of 3.105. All the elements tested were made under the same conditions and using the same types of materials and labour.

The coefficients of variation of the rupture load of the reinforced prisms and reinforced wallettes were very similar and relatively low, showing a greater uniformity of the final load. This fact can be explained by the presence of the steel meshes interlocked by connectors inside the mortars.

Table 3.4 Prism load ratios

Typology	3-blocks prisms/Wallettes	Relation of areas	2-blocks prisms/Wallettes	Relation of areas	3-blocks/2-blocks prisms
Uncoated prisms	0.17	0.53	0.17	0.53	0.98
Prisms with mix ratio 1:3	0.21	0.65	X	X	X
Prisms with a coating of 1.5 cm and mix ratio 1:2:9	0.30	0.93	X	X	X
Prisms with a coating of 3 cm and mix ratio 1:2:9	0.29	0.90	0.32	0.99	0.89
Prisms with a coating of 3 cm of mix ratio 1:1:6	0.31	0.96	0.27	0.84	1.16
Prisms with a coating of 3 cm of mix ratio 1:0.5:4.5	0.23	0.71	X	X	X
Prisms with a coating of 1.5 cm and mix ratio 1:2:9 reinforced	0.29	0.90	X	X	X
Prisms with a coating of 3 cm and mix ratio 1:2:9 reinforced	0.27	0.84	X	X	X
Prisms with a coating of 3 cm and mix ratio 1:1:6 reinforced	0.26	0.81	0.29	0.90	0.90
Average	0.26	0.81	0.26	0.81	0.98
Standard deviation	0.05	0.16	0.06	0.19	0.12
Cov (%)	19	19	24	24	13

According to the Brazilian standard NBR 10837 (2000), the calculation for the admissible load of a masonry wall, P_{Wall}, is given by:

$$P_{Wall} = f_d \cdot A \cdot \left[1 - \left(\frac{h}{40\,t} \right)^3 \right] \quad \text{with } f_d = \frac{f_k}{\gamma_m} = 0.7 \cdot f_{pk} \qquad (3.1)$$

where f_d is the design compressive strength of masonry, A is the resistant cross section area, f_k is the characteristic compressive strength of masonry, γ_m is the partial safety factor for masonry, f_{pk} is the characteristic compressive strength of prism, h and t are the wallet height and thickness, respectively.

In order to obtain a relation between the P_{Wall} and P_{Prism}, not taking into account the safety coefficient and considering the length relation of the wallettes/prism equal to $59/19 = 3.105$, the following relation was obtained: $P_{Wall} / P_{Prism} = 0.344$.

On the other hand, a mean load ratio of the two and three block prisms was obtained experimentally when compared to the results of the wallettes being of the same order of magnitude: 0.26. This behaviour shows that there was no significant difference in the relation between the prisms with 2 or 3 blocks and the wallettes.

The coefficients of variation associate to the rupture load of the reinforced prisms and reinforced wallettes were very similar and relatively low, showing a greater uniformity of the final load. This fact can be explained by the presence of the steel meshes interlocked by connectors inside the mortars.

In literature it is possible to find several equations that relate masonry strength (R_M), mortar strength (R_m) and brick strength (R_l), as the Graf equation (Fontana Cabezas 2015):

$$R_M \left(\text{daN/cm}^2 \right) = \frac{R_l (4 + 0.10 R_m)}{16 + 3 \left(\frac{h}{t} \right)} + K \qquad (3.2)$$

where h is the height of the specimen, t is the thickness of the specimen and K is a constant equal to 10 kg/cm², for a well-executed wall, with 1 cm thick mortar joints. Graf equation was used setting the R_m value as the mean of the average strength obtained for mortars 65 daN/cm² (6.5 MPa, see Table 2.7), R_l equal to 21 daN/cm² (2.05 MPa, see Table 2.2) and different values of the ratio (h/t).

The results for uncoated prisms, using Eq. (3.2), show a ratio of R_M/R_M-experimental equal to approximately 0.33 (0.32 for 2-P1 and 0.34 for 3-P1), a value in accordance with the standard NBR 15961-1 (2011) and the result presented above ($P_{Wall} / P_{Prism} = 0.344$). For coated prisms the ratio R_M/R_M-experimental obtained was approximately 0.97 (1.03 for 3P-3, 1.05 for 3P-4 and 0.88 for 3P-5).

Related to uncoated wallets, the ratio of R_M/R_M-experimental was equal to 0.76 (for W-1), a value in accordance with the standard NBR 15961-1 (2011) and Fontana Cabezas (2015) who showed a result of 0.78. For coated wallets the ratio R_M/R_M-experimental obtained was approximately 1.20 (1.22 for W-3, 1.24 for W-4 and 1.14 for W-5).

(a)

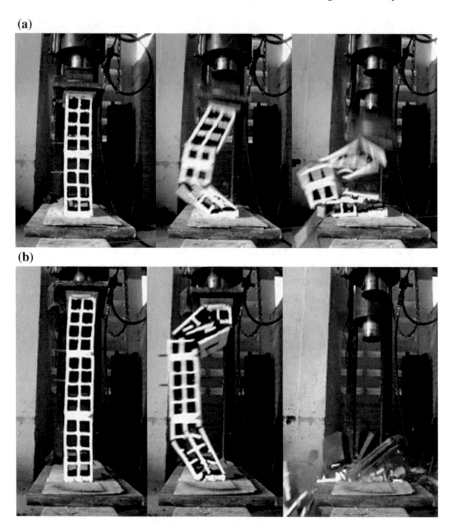

(b)

Fig. 3.2 Rupture of the **a** 2 block prisms and **b** 3 block prisms without coating

Figure 3.2 show the usual type of rupture observed with uncoated prisms, which was abrupt. It is also possible to observe that the prims with 2 or 3 blocks presents the same type of rupture.

(a)

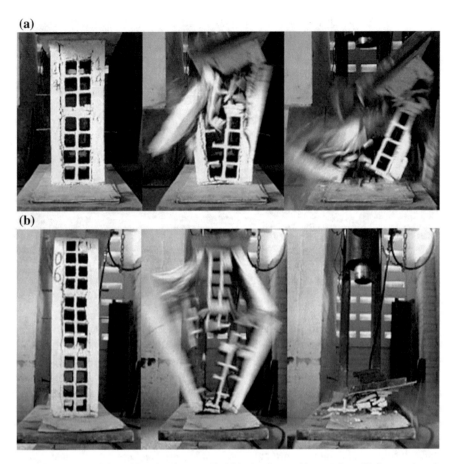

(b)

Fig. 3.3 Rupture of the **a** 2 block prisms and **b** 3 block prisms with a coating of 3.0 cm and a mix ratio of 1:2:9

Figures 3.3 and 3.4 show that the ruptures of the coated prisms start at the septa and it was transferred to the coating layer. This type of rupture it was not abrupt but by shear. The coated prisms present the same type of rupture for 2 or 3 blocks (Fig. 3.5).

Figure 3.6 shows the rupture type observed in 2 block prisms and 3 block prisms with a coating of 3.0 cm and a mix ratio of 1:1:6. It was possible to observe that the rupture in the reinforced prisms was transferred from the septa to the first layer and then to the second layer. This rupture type was less explosive, continuing to be abrupt (Figs. 3.7, 3.8 and 3.9).

Fig. 3.4 Rupture of the 3 block prisms with a mix ratio of 1:3

Finally, Figs. 3.10 and 3.11 show the mean displacements observed during the application of load in the prisms of two and three blocks, respectively. Table 5.5 presents the individual results of the displacements of each type of prism tested in this research. From the analyse of these figures it is possible to observe that for the prisms of 2 and 3 blocks, without coating, the hardness was lesser than in the other prisms studied. In general, it was not observed significant difference behaviour in the prisms made with strong and weak mortar, both for prisms with 2 and 3 blocks. However, the reinforced prisms present higher performance in comparison with the other prisms studied; aspects that suggest the importance of the reinforcement performed.

The main conclusions of this research are: (a) The experimental results showed that the coatings contributed to increasing the vertical compressive resistance of the masonry elements studied; (b) Several types of rupture were observed in the prisms, and it is not possible to define a typical rupture form. On the other hand, the lateral detachment ruptures of the coating layers were frequent; (c) The increase observed in the load related to the reinforcement of the coated two-blocks prisms was approximately 165% and for three-blocks prisms it was of 210%; (d) The relation between the maximum loads of failure of the wallettes and prisms was 0.81; and (e) The ratio of the maximum average loads of 2 block prisms and 3 block prisms was near 1, showing to be equivalent two-blocks and three block prisms.

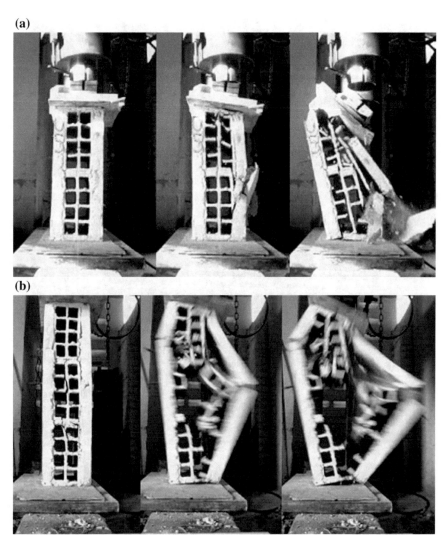

Fig. 3.5 Rupture of the **a** 2 block prisms and **b** 3 block prisms with a coating of 3.0 cm and a mix ratio of 1:1:6

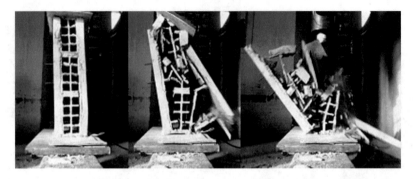

Fig. 3.6 Rupture of the 3 block prisms with a coating of 3.0 cm and a mix ratio of 1:0.5:4.5

(a)

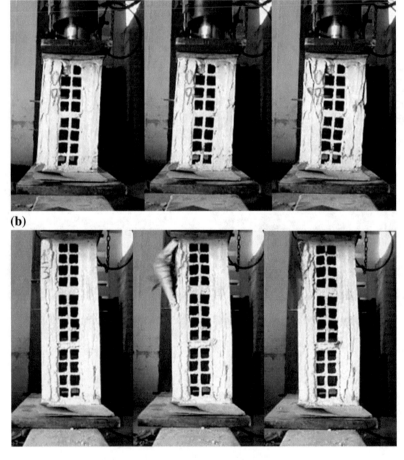

(b)

Fig. 3.7 Rupture of the **a** 2 block prisms and **b** 3 block prisms with a coating of 3.0 cm and a mix ratio of 1:1:6 reinforced with a POP mesh of 10 × 10 cm and 4.2 mm of diameter

Fig. 3.8 Rupture of the 3 block prisms with a coating of 1.5 cm and a mix ratio of 1:2:9 reinforced with a POP mesh of 10 × 10 cm and 4.2 mm of diameter

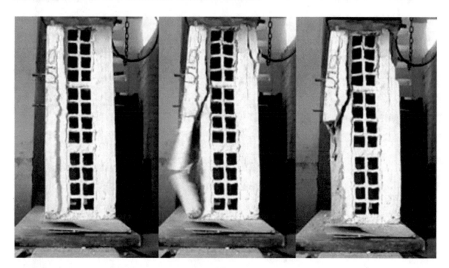

Fig. 3.9 Rupture of the 3 block prisms with a coating of 3 cm and a mix ratio of 1:1:6 reinforced with a POP mesh of 10 × 10 cm and 4.2 mm of diameter

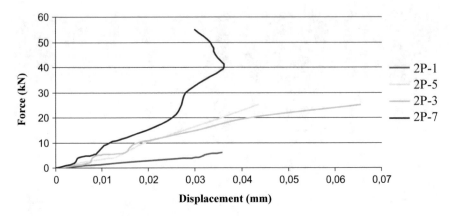

Fig. 3.10   Load versus displacement for prisms of 2 blocks

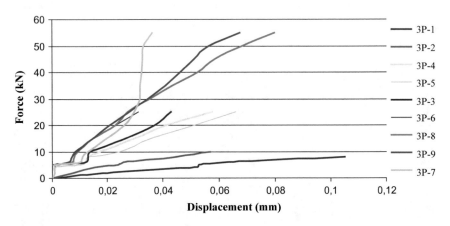

Fig. 3.11   Load versus displacement for prisms of 3 blocks

References

J.J. Fontana Cabezas, Mechanical properties of load bearing walls made of Uruguayan hollow ceramic bricks. Revista de la Construcción 14(3), 15–21 (2015)

NBR 10837, *Hollow Concrete Blocks—Bases for Design of Structural Masonry Procedure*, Rio de Janeiro, Brazil (2000)

NBR 15961–1, *Structural Masonry—Concrete Blocks—Part 1: Design*, Rio de Janeiro, Brazil (2011)

Chapter 4
Structural Performance of Resistant Masonry Elements

4.1 Materials

In order to evaluate the behaviour of resistant masonry elements used for structural purposes, blocks, prisms, wallets, walls and foundation elements experimentally tested. The prisms were made and tested in the Materials Laboratory of the Catholic University while the walls, walls and foundation elements were made and tested in the ITEP. The details of the elements and tests performed are presented below.

4.1.1 Concrete and Ceramic Blocks

Ceramic and concrete blocks used in the research were of the same type as those usually employed in real resistant masonry buildings in the region. The dimensional characteristics of these blocks were obtained through tests of 60 ceramic blocks and 30 concrete blocks. Table 4.1 summarizes the results obtained for both type of blocks studied.

4.1.2 Sand and Mortars

The sand used in the preparation of the mortars for laying and coating the tested models is usually found in the MRR and all the lot used in the development of the research was acquired from the same supplier. Table 4.2 summarizes the results of sand characterization.

The mortars used both in the laying of the blocks and in the coating were defined from cement, lime and sand mixtures in proportions of 1:2:9, 1:1:6 and 1:0.5:4.5 by volume.

© The Author(s), under exclusive license to Springer Nature Switzerland AG 2019   37
J. M. P. Q. Delgado et al., *Structural Performance of Masonry Elements*, SpringerBriefs
in Applied Sciences and Technology, https://doi.org/10.1007/978-3-030-03270-8_4

Table 4.1 Mean characteristics of the blocks testes

(a) **Ceramic blocks**—NBR 15270-1 (2005)

Length (mm)	190
Width (mm)	90
Height (mm)	190
Thickness of horizontal and vertical septa (mm)	7.0
Compression strength (MPa)	**2.15**

(b) **Concrete blocks**—NBR 6136 (2014)

Length (mm)	390
Width (mm)	90
Height (mm)	190
Thickness of horizontal and vertical septa (mm)	21.5
Thickness of the internal transverse septa (mm)	22.5
Thickness of external transverse septa (mm)	25.0
Compression strength (MPa)	**2.30**

Table 4.2 Characteristics of the natural sand used

Maximum characteristic size (mm)—NBR 7211 (2009)	4.80
Fineness module—NBR NM 248 (2003)	3.20
Unit mass (g/cm^3)—NBR NM 45 (2006)	1.42
Specific mass (g/cm^3)—NBR 9776 (1987)	2.60
Swelling—NBR 6467 (2009)	1.25
Critical humidity (%)—NBR 6467 (2009)	3.00
Powdery material content (%)—NBR 7219 (1982)	1.26

Table 4.3 Characterization of mortars used

Item	Mean value		
	1:2:9	1:1:6	1:0.5:4.5
Compressive strength—MPa	4.00	5.80	6.23

Table 4.3 presents the values of the compressive strength of the mortars, obtained through tests of 15 specimens in accordance with the Brazilian standards NBR 7215 (1994) and NBR 7222 (2011). The amount of cement used in the mortars was 220 and 150 kg/m^3 for the mixture proportions of 1:1:6 and 1:2:9, respectively.

In the case of concrete blocks, the laying grout cords were applied both to the longitudinal and to the transverse septa of the blocks, a situation that is usually referred to as a total settlement.

4.1.3   Steel Mesh and Connectors

Two types of steel mesh were used as reinforcement of mortar coating in the prisms tested: one using galvanized steel and other with ribbed welded steel. The galvanized steel mesh is formed of wires with a diameter of 2.7 mm and a spacing of 5 cm in the horizontal direction and 10 cm in the vertical, making a steel area of 1.06 and 0.53 cm^2/m, respectively. The ribbed welded steel mesh had wires with a diameter of 4.2 mm and a spacing of 10 cm in the horizontal and vertical directions, making a steel section of 1.38 cm^2/m in both directions. Steel connectors were 5.0 mm in diameter.

4.1.4   Prisms

Approximately 500 prisms made with three blocks in vertical direction were tested—300 made with ceramic blocks and 200 with concrete blocks. The prisms were all capped at the top and bottom with cement paste in a thickness of 5 mm to obtain a uniform surface. Prisms were made in order to reproduce the conditions found in the daily practice of resistant masonry constructions in the region. The typology of prisms tested together with the corresponding acronym to identify each pattern in lab tests is presented below.

- Prisms of uncoated concrete and ceramic blocks (PSR);
- Prisms of concrete and ceramic blocks coated with 3.0 cm of mortar coating (PR30MM);
- Prisms of concrete and ceramic blocks coated with 3.0 cm of reinforced mortar (PCRTP—with ribbed welded steel—and PCRTG—with galvanized steel);
- Prisms of concrete and ceramic blocks coated with 3.0 cm of reinforced mortar with steel connectors (PCRTP-C and PCRTG-C);
- Prisms of concrete and ceramic blocks coated with 3.0 cm of mortar and an additional reinforced mortar layer with steel connectors (PRAATG–C e PRAATP–C).

All prisms were initially coated with a 5 mm layer of a scratch coating in a mixture proportion of 1:3 (cement and sand) and after 24 h they received an additional coating layer of 2.5 cm in a mixture proportion of 1:1:6 (cement, lime, and sand) by volume. The prisms were submitted to a curing process under ambient conditions for a period of at least 28 days.

The execution of the coating of the prisms with a layer of 3.0 cm was carried out in four stages, as described below:

- Apply the scratch coating;
- Apply one layer of mortar (1.0 cm thick);
- Apply and install of steel meshes with connectors and
- Apply a second layer of 1.5 cm thick of mortar coating 1.5 cm leaving steel meshes fully immersed in the mortar.

The prisms that received a reinforced mortar layer over the existing unreinforced coating were initially made following the same procedure used in the prisms with mortar coating of 3.0 cm in thickness without steel meshes, which were coated in a single step by means of jigs wooden.

Once this step is completed and after a curing period of 28 days, transverse holes were made in the prisms through which steel connectors were inserted to install the steel meshes on the surface of the coating. Completed this operation, the second layer of coating mortar was applied over the steel meshes, leaving it fully involved and creating a final 6.0 cm thickness mortar layer. All prisms were capped with cement paste at the top and bottom. The transport of the specimens to the test machine required special care in order to avoid damages.

4.1.5   Wallets Specimens

The wallet is an element that better represent a real masonry wall because it contains all its parts, i.e.: bed and head joints and individual units lay in and bound together by mortar.

In order to analyse the influence of the mortar mixture proportions, the coating thickness and the reinforcement with steel meshes interlinked by connectors, 154 ceramic wallets were made. The specimens had dimensions of $0.09 \times 0.60 \times 1.20\,\text{m}^3$. Ceramic blocks of eight holes with dimensions of $9 \times 19 \times 19\,\text{cm}^3$, with bed and head mortar joints made from a mixture proportion of 1:1:6 (cement, lime and sand) in volume. Figure 4.1 shows the types of tested wallets and Table 4.4 presents the characteristics of the wallets tested, all made with bed and head joints made with

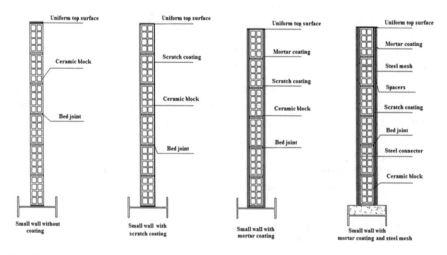

Fig. 4.1 Typical schemes of the tested walls

Table 4.4 Wallets characteristics

Ref.	Scratch coating mixture proportion	Mortar coating mixture proportion	Mortar thickness on each side (cm)	Obs.
1	–	–	–	w/o coating
2	1:3	–	–	Only w/scratch coating
3	1:3	1:2:9	1.5	w/o steel mesh
4	1:3	1:2:9	3.0	w/o steel mesh
5	1:3	1:1:6	3.0	w/o steel mesh
6	1:3	1:0.5:4.5	3.0	w/o steel mesh
7	1:3	1:1:6	3.0	w/o steel mesh w/additive
8	1:3	1:2:9	3.0	w/steel mesh
9	1:3	1:2:9	1.5	w/steel mesh
10	1:3	1:1:6	3.0	w/steel mesh

Fig. 4.2 Wallets construction steps—elevation, scratch coating and mortar coating application

mortar of mixture proportion of 1:1:6 in volume. Fifteen specimens from each type were constructed.

The wallets were made using a three-course stage by day with bed and head mortar joints of 1.0 cm in thickness. They were built over steel 8 inches—H channel section, some of them filled of concrete to facilitate to apply mortar coating to the faces of the wall after the installation of the steel meshes of the same type of those used in the prisms. Figure 4.2 shows the steps of confection of the wallets tested and Fig. 4.3 exhibits the process of installation of steel meshes and connectors before the application of mortar coating layer, also showed in the figure (Andrade 2007).

Special devices to transport the specimens and perform the works to smooth the top and bottom surface of the walls were created as it can be seen in Fig. 4.4.

The tests were carried out on a steel reaction frame with digitally controlled hydraulic loading system and digital data acquisition system. The vertical load was

Fig. 4.3 Steps of drilling, installation of connectors, steel mesh, mortar coating and spacers

Fig. 4.4 Details of the handling the specimen and execution of top bottom surface

applied using hydraulic jacks with 200 mm maximum stroke and 1,500 kN of compression capacity. The applying loading velocity used in the tests was 0.05 MPa/s.

4.1.6 Steel Mesh and Connector

Two types of steel mesh were used as reinforcement of mortar coating in the prisms tested: one using galvanized steel and other with ribbed welded steel. The galvanized steel mesh is formed of wires with a diameter of 2.7 mm and a spacing of 5 cm in the horizontal direction and 10 cm in the vertical, making a steel area of 1.06 and 0.53 cm^2/m, respectively. The ribbed welded steel mesh had wires with a diameter of 4.2 mm and a spacing of 10 cm in the horizontal and vertical directions, making a steel section of 1.38 cm^2/m in both directions. Steel connectors were 5.0 mm in diameter.

Fig. 4.5 Prisms of uncoated ceramic blocks: evolution of rupture

Fig. 4.6 Ceramic block prisms with coating: evolution of rupture

4.2 Results and Discussion

The results from the performed tests are presented and discussed in the following sections. Figures 4.5, 4.6, 4.7 and 4.8 illustrate the rupture modes observed in the tested ceramic block prisms. As it can be observed, the ruptures are fragile and explosive, characterized by an immediate loss of the system's strength capacity soon after reaching the maximum load. For the coated prisms, it was observed that the cracking process starts in the horizontal septa of the blocks and from this moment, the two coating layers primarily carry the load. It should be noted, however, that several ways of rupture were observed, and it is a hard task to choose one that represents the universe of prisms tested. Several factors can influence the rupture process, such as the quality of the workmanship used to construct the specimens, the thickness, and uniformity of the bed mortar joints, among others. Nevertheless, the following ruptures modes can be highlighted as more frequent (Mota 2006):

– Coating cover rupture caused by excessive lateral displacement of bed mortar joints (Fig. 4.6, third photo);
– Rupture by a detachment of the coating layers;
– Buckling rupture of mortar coating layers without connectors (Fig. 4.7, third photo);

Table 4.5 summarizes the rupture load of ceramic block prisms.

Fig. 4.7 Ceramic block prisms with coating and steel mesh

Fig. 4.8 Ceramic block prisms with a reinforced mortar coating layer with connectors

As shown, the increase in load carrying capacity of the prism coating is significant. When the uncoated prism is compared to the uncoated prism, it is observed an increase of approximately 33% in its load capacity. When one compares the values obtained for the prism coated with that from the prism without any coating, there is an increase in the average of rupture load of approximately 130% that consubstantiates a significant increase in the load capacity generated by the mortar coating. The increment obtained with the installation of steel mesh inside the mortar coating was of 31 and 24%, for ribbed welded steel mesh and galvanized mesh, respectively.

Table 4.5 Rupture loads of ceramic block prisms—Blocks with fbk = 2.15 MPa

Specimen	Average rupture load (kN)	Dispersion measures		Characteristic rupture load (kN)
		Standard deviation (kN)	Cov (%)	
Uncoated prisms (PSR)	32.46	6.52	20.11	21.70
Prisms with scratch coating only (5 mm)	43.35	7.35	21.25	31.22
Prisms with 30 mm mortar coating (PR30MM)	74.93	13.90	18.55	52.00
Prisms coated with 3.0 cm of reinforced mortar with ribbed welded steel (PRCTP)	98.07	19.46	19.84	65.96
Prisms coated with 3.0 cm of reinforced mortar with galvanized steel (PRCTG)	92.86	21.59	23.25	57.23
Prisms coated with 3.0 cm of mortar and an additional reinforced mortar layer with ribbed welded steel and connectors (PRAATP–C)	205.44	16.06	7.82	178.93
Prisms coated with 3.0 cm of mortar and an additional reinforced mortar layer with galvanized steel and connectors (PRAATP–C)	150.18	25.33	16.86	108.39

The rupture observed in the prisms tested was always abrupt, for both coated and uncoated prisms. The rupture occurred due to the failure of the horizontal septa of the blocks by transverse tensile deformation followed by the collapse of the septa corresponding to the joints of mortar, causing the loss of equilibrium of the specimen. The coated prisms, reinforced with steel mesh and connectors also showed abrupt rupture, however with compressive loads well above the others. This is due to the presence of the connectors and their action of preventing the horizontal displacements of the septa.

The models with the highest average load of rupture, with a lower coefficient of variation, were those reinforced with ribbed welded steel and connectors.

Analyses with varying thicknesses and mixture proportions of the mortar were additionally performed and the results obtained confirmed the above-mentioned observations. Further details can be found in Oliveira et al. (2011).

Table 4.6 Rupture loads of concrete block prisms—Blocks with fbk = 2.30 MPa

Specimen	Average rupture load (kN)	Dispersion measures		Characteristic rupture load (kN)
		Standard deviation (kN)	Cov (%)	
Uncoated prisms (PSR)	86.22	11.47	13.30	67.30
Prisms with 30 mm mortar coating (PCR)	148.23	19.89	13.42	115.41
Prisms coated with 3.0 cm of reinforced mortar with ribbed welded steel (PRCTP)	223.43	17.07	7.64	195.26
Prisms coated with 3.0 cm of reinforced mortar with galvanized steel (PRCTG)	187.30	14.11	7.53	164.02
Prisms coated with 3.0 cm of reinforced mortar with ribbed welded steel and connectors (PCRTP-C)	251.70	36.43	14.47	191.59
Prisms coated with 3.0 cm of reinforced mortar with galvanized steel and connectors (PCRTG-C)	233.36	32.61	13.97	179.56
Prisms coated with 3.0 cm of mortar and additional reinforced mortar layer with ribbed welded steel and connectors (PRAATP–C)	371.36	38.37	10.33	308.06
Prisms coated with 3.0 cm of mortar and an additional reinforced mortar layer with galvanized steel and connectors (PRAATG–C)	392.42	34.80	8.87	335.03

4.2.1 Concrete Brick Prisms

Table 4.6 summarizes the rupture load of concrete block prisms. As shown, the increase in load capacity due to the prism coating was significant. In fact, in the case coating without steel mesh, the increase in the average failure load, when compared to the uncoated prism, was approximately 72% whereas in the case of the coating with steel mesh inside the increase reached 159 and 117%, for ribbed welded and galvanized steel meshes, respectively. It was also possible to observe that the existence of the steel mesh inside the coating concurred to an increase in the load capacity of the coated prism. In the case of ribbed welded steel mesh, the average increase was approximately 51% and in the case of galvanized steel mesh, this increase was 27%. This behaviour indicates an important participation of the steel mesh that can be exploited for possible retrofitting works of real masonry building. Figures 4.9, 4.10, 4.11, 4.12 and 4.13 shown the rupture modes observed for the uncoated and coated concrete block prisms (Araújo Neto 2006; Azevedo 2010).

Fig. 4.9 Rupture of uncoated concrete block prisms

Fig. 4.10 Rupture of concrete block prisms with a coating without steel mesh

Fig. 4.11 Rupture of concrete block prisms with coating and steel mesh

In the case of uncoated concrete block prisms, the cracks observed were located on the faces of the blocks and on the settlement surface. The former presented a markedly random characteristic, while the latter presented a pattern with more regularity.

The rupture observed was less abrupt than that registered in the uncoated prisms made with ceramic blocks and in some of the prisms, displacements of the walls of the blocks were observed, as it can be seen in Fig. 4.7.

For the coated prisms, without the addition of steel meshes inside the mortar coating, the most frequent rupture mode is indicated in Fig. 4.8. This type of rupture was characterized by a cracking process located on the front and back faces of the blocks, suggesting a rupture generated by transversal tensile strains.

Fig. 4.12 Rupture of concrete block prisms with a coating, steel mesh, and connectors

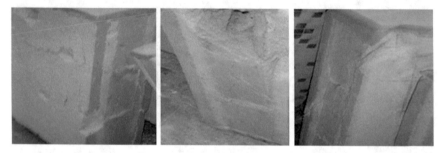

Fig. 4.13 Rupture of concrete block prisms coated with 3.0 cm of mortar and an additional reinforced mortar layer

For the prisms with steel meshes inside the mortar coating, the most frequent failure mode is shown in Fig. 4.11. It was similar to that observed for coated prisms without steel meshes. It deserves attention, however, the improvement in the performance of the prisms generated by the steel meshes inside the mortar coating and the fact that was observed a slightly change in the failure mode that became less abrupt than that observed in prisms with the unreinforced coating.

Another aspect that should be emphasized was the greater regularity in the value of the rupture load of the coated prisms with steel mesh that was not present in the prisms with a coating but without steel mesh inside. This more regular behaviour can be attributed exclusively to the existence of steel meshes inside the mortar coatings. It is possible to infer that the coating with reinforced mortars should have contributed to the load capacity of the prisms since the beginning of the loading process, but their more effective participation occurred when the blocks were no longer able to withstand the applied load.

Table 4.7 Comparison between failure loads of concrete and ceramic block prisms

Comparison indicator	Type of block	
	Concrete	Ceramic
Load increase after application of mortar coating (%)	71.91	130.84
Load increase after the application of steel mesh without connectors (%)	50.74	30.88
Load increase after the application of the connectors to the steel mesh (%)	69.81	–
Load increase after the application of an additional reinforced mortar layer with connectors over existing mortar coating (%)	164.76	174.18

When the steel meshes were installed inside the mortar coating, it was observed an increase in the load capacity of the coated prism of approximately 27% in the case of a galvanized steel mesh and 51% in the case of ribbed welded steel mesh. This better performance of ribbed welded steel mesh can be attributed to its better adhesion conditions.

The coating layer contributed significantly to the increase of the failure load of the prisms. The increment obtained with the installation of steel meshes inside the mortar coating also provided a significant improvement in the loading capacity of the studied prisms around 65%. Table 4.7 presents a comparison of the results of the concrete block and ceramic block prisms.

Analysing data in Table 4.7, it is possible to observe the influence of the mortar coating on the ceramic block prisms was greater than on the concrete block one. The steel mesh provides more than 50% increase in failure load of concrete block prisms and a lower influence on the ceramic block prisms—30.88%. This increase in load capacity of ceramic block prisms can be attributed to the high strength of the ceramic when compared to the strength of the block itself. In the course of the tests, it was possible to observe that the ceramic blocks prisms collapsed before the prism reached the failure load and, consequently, the mortar coating was much more demanded. The lower contribution of the steel mesh to the increase in the load capacity of the ceramic block prisms can be explained by the fact that the steel mesh failed in buckling when the critical load was reached, taking, this way, the mortar coating layer together. The action of the connectors proved to be fundamental to combat the transverse tensile strains that motivate the rupture and the provision of some ductility to the specimen. For the concrete blocks prisms, the buckling of the steel meshes was partially counteracted by the connectors that interconnect the two steel meshes immersed in the mortar coating layers.

4.2.2 Wallet Specimens

Tables 4.8, 4.9 and 4.10 show the results of compressive tests in wallets specimens. The rupture mode of most of the wallet specimens occurred in the septum of the blocks in the upper region, close to the point of application of the load, and later they were accompanied by cracks in the interface of mortar coating and the scratch coating layers. Figure 4.14 shows details of the characteristic rupture mode of coated wallet specimens.

This behaviour is due to the tri-axial stress state to which the bed mortar joint is subjected as a consequence of its confinement between the blocks. This stress state generates horizontal tensile stresses to horizontal septa of the blocks, due to the mobilized its adhesion between with bed mortar joints. Thus, when tensile stress exceeds the tensile strength of these septa, they crack, distributing such stress to the others septa and to the mortar coating that tends to crack or detach, if there is no satisfactory adhesion, before the rupture of the specimen.

Table 4.8 Loads corresponding to first crack in the blocks of wallet specimens

Small walls	Average load of the first crack (kN)	Dispersion measures		First crack characteristic load (kN)
		Standard deviation (kN)	Cov (%)	
Uncoated specimen	41.40	11.60	27.90	22.30
Scratch coating specimen	67.90	19.60	28.80	35.60
Coating with mortar mixture proportion of 1:2:9 and thickness of 1.5 cm specimen	85.30	22.70	26.60	47.80
Coating with mortar mixture proportion of 1:2:9 and thickness of 3.0 cm specimen	96.10	34.50	35.90	39.20
Coating with mortar mixture proportion of 1:1:6 and thickness of 3.0 cm specimen	105.50	43.40	41.10	33.90
Coating with mortar mixture proportion of 1:0.5:4.5 and thickness of 3.0 cm specimen	170.30	54.20	31.80	80.90
Coating with mortar mixture proportion of 1:2:9 and thickness of 1.5 cm + Reinforced mortar coating of 3.0 thickness layer with mixture proportion of 1:1:6 specimen	242.00	84.60	35.00	102.40
Coating with mortar mixture proportion of 1:2:9 and thickness of 3.0 cm + Reinforced mortar coating of 3.0 thickness layer with mixture proportion of 1:1:6 specimen	254.80	57.50	22.60	159.90

Table 4.9 Loads corresponding to first crack in mortar coating of wallet specimens

Small walls	Average load of the first crack (kN)	Dispersion measures		First crack characteristic load (kN)
		Standard deviation kN)	Cov (%)	
Uncoated specimen	–	–	–	–
Scratch coating specimen	–	–	–	–
Coating with mortar mixture proportion of 1:2:9 and thickness of 1.5 cm specimen	–	–	–	–
Coating with mortar mixture proportion of 1:2:9 and thickness of 3.0 cm specimen	123.70	34.00	27.50	67.60
Coating with mortar mixture proportion of 1:1:6 and thickness of 3.0 cm specimen	133.10	48.70	36.30	52.70
Coating with mortar mixture proportion of 1:0.5:4.5 and thickness of 3.0 cm specimen	240.80	42.60	17.70	170.50
Coating with mortar mixture proportion of 1:2:9 and thickness of 1.5 cm + Reinforced mortar coating of 3.0 thickness layer with mixture proportion of 1:1:6 specimen	252.10	44.90	17.80	178.00
Coating with mortar mixture proportion of 1:2:9 and thickness of 3.0 cm + Reinforced mortar coating of 3.0 thickness layer with mixture proportion of 1:1:6 specimen	206.60	59.70	28.80	108.10

In the case of wallet specimens without reinforced mortar coating layer, cracks in the septa of the blocks occurred before cracks were observed in the mortar coating mortar, an aspect that indicates an effective participation of the coating in the compressive behaviour of the specimen. For wallet specimens with reinforced mortar coating with the connector, the initial crack occurred at the interface between mortar coating mortar and reinforcement mortar. This happened possibly due to the greater deformability of reinforced mortar layer and its positioning (without confinement) in relation to the core of the coated and confined specimens associated with the lower adhesion at the interface between the old and the new coating.

Taking into account the results of wallet specimens tested it is possible to formulate the following considerations.

Table 4.10 Failure load of wallet specimens

Small walls	Average failure load (kN)	Dispersion measures		Characteristic failure load (kN)
		Standard deviation (kN)	Cov (%)	
Uncoated specimen	56.30	8.70	15.40	41.90
Scratch coating specimen	84.90	16.30	19.20	58.00
Coating with mortar mixture proportion of 1:2:9 and thickness of 1.5 cm specimen	130.40	24.40	18.70	90.10
Coating with mortar mixture proportion of 1:2:9 and thickness of 3.0 cm specimen	156.50	16.10	10.30	129.90
Coating with mortar mixture proportion of 1:1:6 and thickness of 3.0 cm specimen	168.30	33.30	19.80	113.40
Coating with mortar mixture proportion of 1:0.5:4.5 and thickness of 3.0 cm specimen	262.20	42.70	16.30	191.70
Coating with mortar mixture proportion of 1:2:9 and thickness of 1.5 cm + Reinforced mortar coating of 3.0 thickness layer with mixture proportion of 1:1:6 specimen	321.00	47.70	14.90	242.30
Coating with mortar mixture proportion of 1:2:9 and thickness of 3.0 cm + Reinforced mortar coating of 3.0 thickness layer with mixture proportion of 1:1:6 specimen	367.00	49.30	13.40	285.70
Coating with mortar mixture proportion of 1:1:6 and thickness of 3.0 cm + Reinforced mortar coating of 3.0 thickness layer with mixture proportion of 1:1:6 specimen	417.09	62.99	15.10	313.35

4.2.3 Coating Influence

It was observed that the simple application of a scratch coating thin layer (5.0 mm of thickness) was capable to generate an average increase of 50.7% in the failure load of the wallet specimens, without, however, changing the sudden form of collapse. Figure 4.15 shows this behaviour.

By analysing the behaviour of the wallet specimens (Figs. 4.15 and 4.16), it was possible to observe an increase of the inclination of the load x displacement curve. The average stiffness of the wallet specimens without scratch coating layer was of 17.28 kN/m, while to the same specimens with scratch coating layer it was of 21.60 kN/m, representing an increase of 25%. In these figures, the stretches of the load-displacement curves located after the values of the maximum loads represent the unloading curve of the press and have no physical meaning.

Fig. 4.14 Rupture mode of coated wallet specimens

The results show that the scratch coating layer generated a 50% increase in the load capacity of the wallet specimens and a 25% increase in its stiffness without, however, changing the abrupt failure mode.

4.2.4 Mortar Coating Mixture Proportion Influence

A comparison of the results of the wallet specimens according the mortar coating mixture proportion (1:2:9, 1:1:6 and 1:0.5:4.6) it was possible to observe a discrete increase in the load capacity of the specimen of about 7.5%, between the 1:2:9 and 1:1:6 mixture proportion and a considerable increase of 55.8%, between mortar mixture proportions of 1:0.5:4.5 and 1:1:6.

By analysing the compressive behaviour of the wallet specimen tested (Figs. 4.17, 4.18 and 4.19) it is possible to observe that there is an increase in the stiffness of the coated wallets with the increase of the cement content in the mortar mixture. The wallet coated with a mortar layer mixture proportion of 1:2:9 showed an average rigidity of 48 kN/m while those coated with 1:1:6 mortar mixture proportion showed an average rigidity of 58 kN/m—an increase of 21%. Wallet specimens coated with mortar layer with mixture proportion of 1:0.5:4.5 presented average rigidity of the order of 64 kN/m, which represents an increase of 10% in relation to the previous one. In these figures, the stretches of the load-displacement curves located after the values of the maximum loads represent the unloading curve of the press and have no physical meaning.

The increase in cement content of mortar coating layer mixture proportions generated increases in the load capacity and stiffness of wallet specimens, without, however, changing the abrupt failure mode.

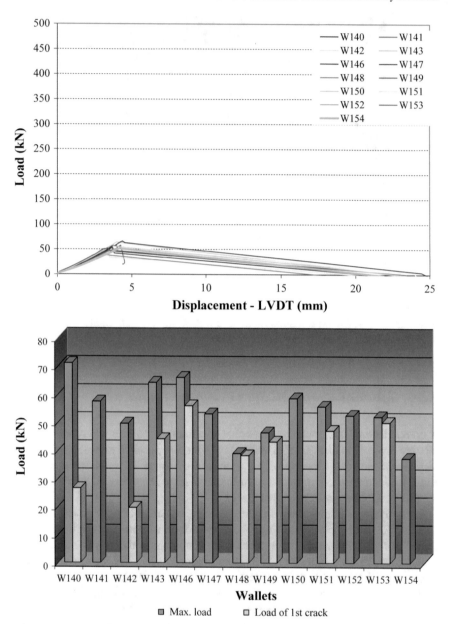

Fig. 4.15 Load versus displacement diagram of wallet specimens without the scratch coating layer

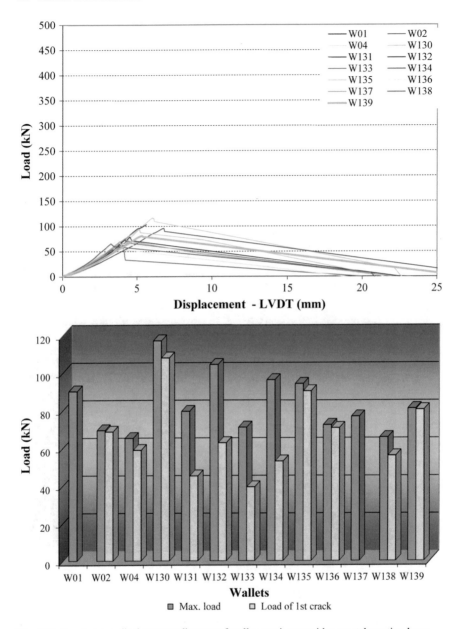

Fig. 4.16 Load versus displacement diagram of wallet specimens with a scratch coating layer

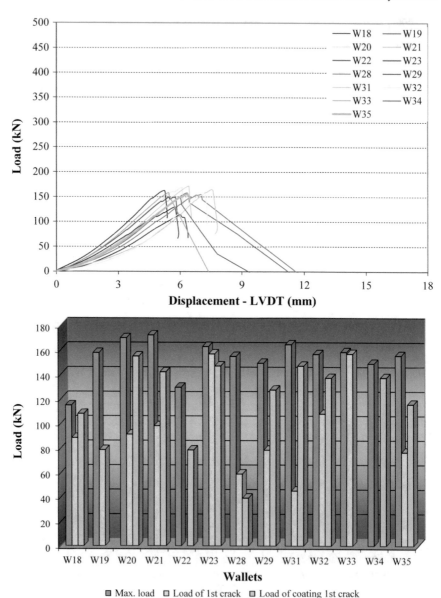

Fig. 4.17 Load versus displacement diagram of wallet specimens coated with a 3.0 cm in thickness mortar layer with a mixture proportion of 1:2:9

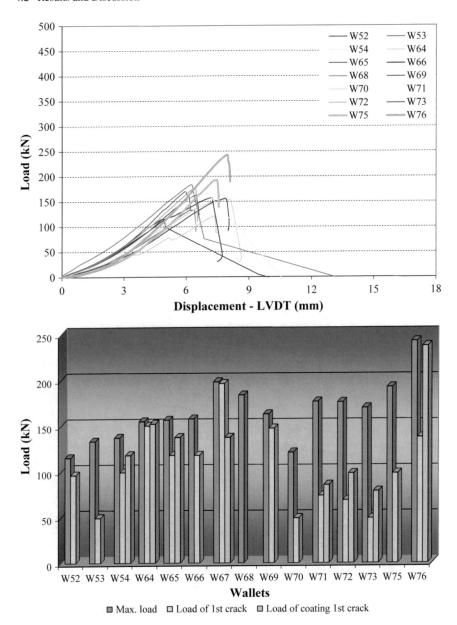

Fig. 4.18 Load versus displacement diagram of wallet specimens coated with a 3.0 cm in thickness mortar layer with a mixture proportion of 1:1:6

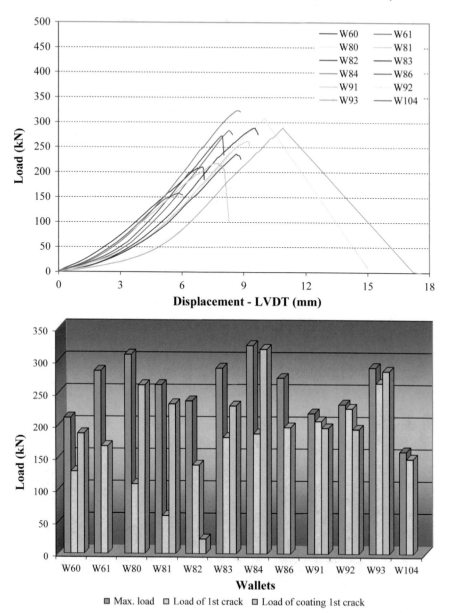

Fig. 4.19 Load versus displacement diagram of wallet specimens coated with a 3.0 cm in thickness mortar layer with a mixture proportion of 1:0.5:4.5

4.2.5 Influence of the Additive

The following Figs. 4.20 and 4.21 show the influence of the additive for a wallet with mixture proportion of 1:2:6 and with thicknesses of 3.0 cm. It is possible to observe that practically no changes were observed.

4.2.6 Mortar Coating Layer Thickness Influence

The following figures promoting a comparative analysis of the average strength of wallets as a function of the thickness of the mortar coating layer, with mixture proportion of 1:2:9 and thicknesses of 1.5 and 3.0 cm (Figures 4.22, 4.23, 4.24 and 4.25).

It can be observed that the failure load increased with the increment of the thickness. It can also be noted that this increase was of about 8.5%, between the thicknesses of 1.5–3.0 cm. The increases in failure loads of wallets, comparing their values with those from the model with only a thin scratch coating layer, were of 58 and 72%, respectively.

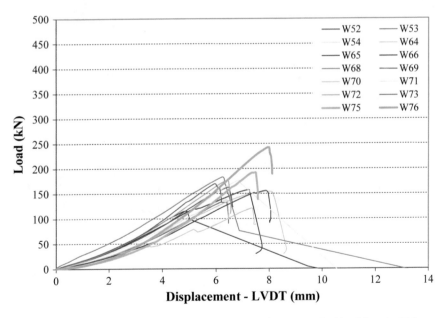

Fig. 4.20 Load versus displacement diagram of wallet specimens coated with a 3.0 cm in thickness mortar layer with a mixture proportion of 1:1:6

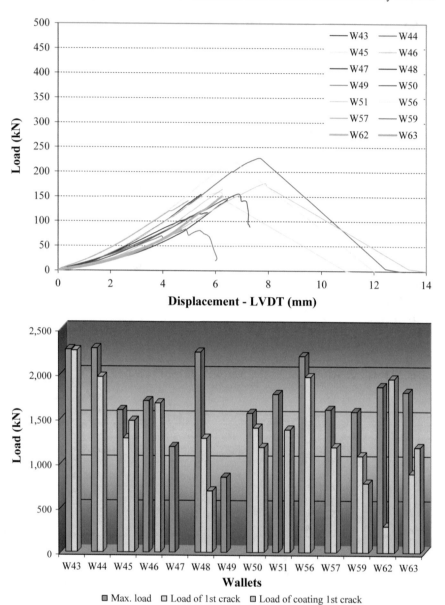

Fig. 4.21 Load versus displacement diagram of wallet specimens coated with a 3.0 cm in thickness mortar layer with a mixture proportion of 1:1:6 and an additive

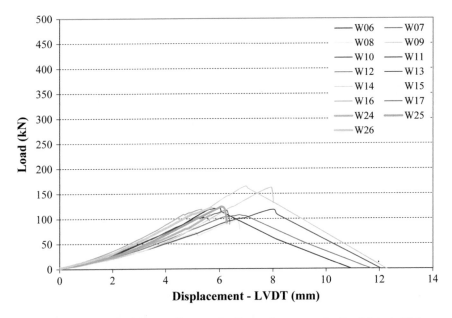

Fig. 4.22 Load versus displacement diagram of wallet specimens coated with a 1.5 cm in thickness mortar layer with a mixture proportion of 1:2:9

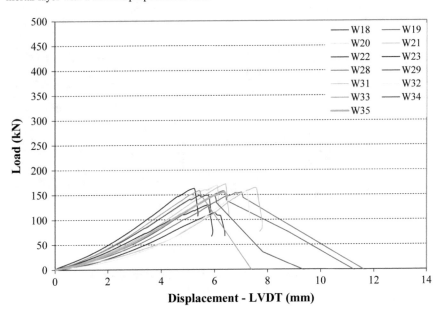

Fig. 4.23 Load versus displacement diagram of wallet specimens coated with a 3.0 cm in thickness mortar layer with a mixture proportion of 1:2:9

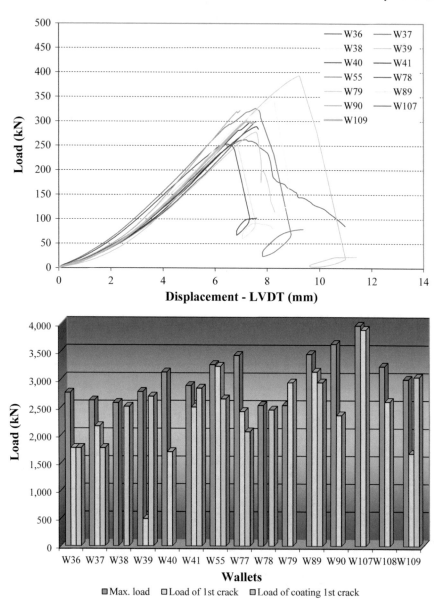

Fig. 4.24 Load versus displacement diagram of wallet specimens coated with a 1.5 cm in thickness mortar layer with a mixture proportion of 1:2:9 with reinforcement

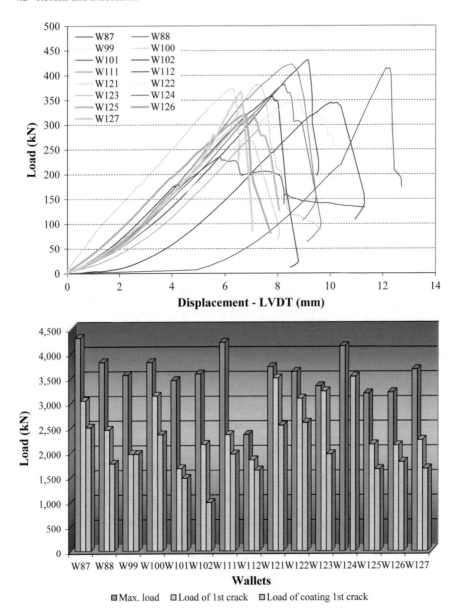

Fig. 4.25 Load versus displacement diagram of wallet specimens coated with a 3.0 cm in thickness mortar layer with a mixture proportion of 1:2:9 with reinforcement

4.2.7 Influence of Reinforced Mortar Layer with Steel Meshes

Wallets specimens with a 3.0 cm thick mortar coating layer in a mixture proportions of 1:2:9 and 1:1:6 were reinforced with steel meshes (10×10 cm^2) with 4.2 diameter wires interconnected through steel connector with 6.0 mm in diameter spaced of 20 cm and covered with a mortar layer with a mixture proportion of 1:1:6.

The results show an expressive increase in the load capacity of the specimens when reinforced with steel meshes interlocked by connectors. Increases of 134.50 and 147.82% were observed in the specimens with a mixture proportion mortar coating of 1:2:9 and 1:1:6, respectively.

Figures 4.26, 4.27, 4.28 and 4.29 show the typical force versus displacement curves of the various typologies of wallets tested. In these figures, the stretches of the load-displacement curves located after the values of the maximum loads represent the unloading curve of the press and have no physical meaning.

Observing the post-peak behaviour of the wallet specimen reinforced with steel meshes, one can note the importance of the connectors interlocking the steel meshes. While the 90° hooks of the steel connectors did not open, the specimens maintained the load capacity similar to that from specimens without reinforcement.

Finally, Fig. 4.30 presents a resume of all experimental results obtained with different wallets.

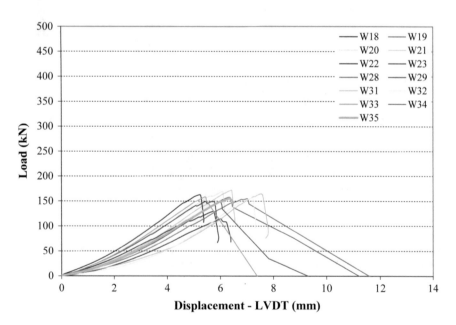

Fig. 4.26 Load versus displacement diagram of wallet specimens coated with a 3.0 cm in thickness mortar layer with a mixture proportion of 1:2:9 without reinforcement

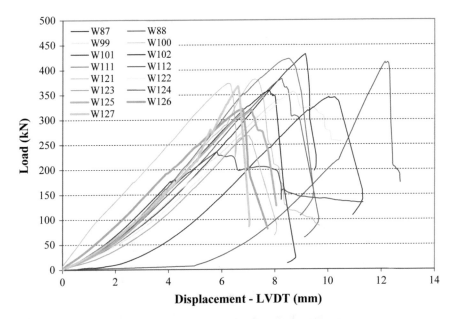

Fig. 4.27 Load versus displacement diagram of wallet specimens coated with a 3.0 cm in thickness mortar layer with a mixture proportion of 1:2:9 with reinforcement

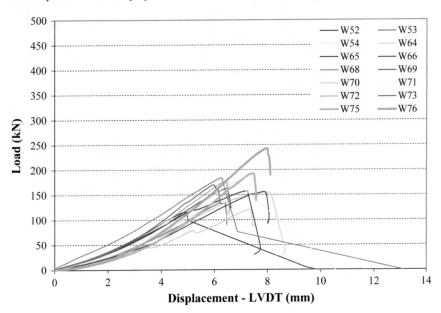

Fig. 4.28 Load versus displacement diagram of wallet specimens coated with a 3.0 cm in thickness mortar layer with a mixture proportion of 1:1:6 without reinforcement

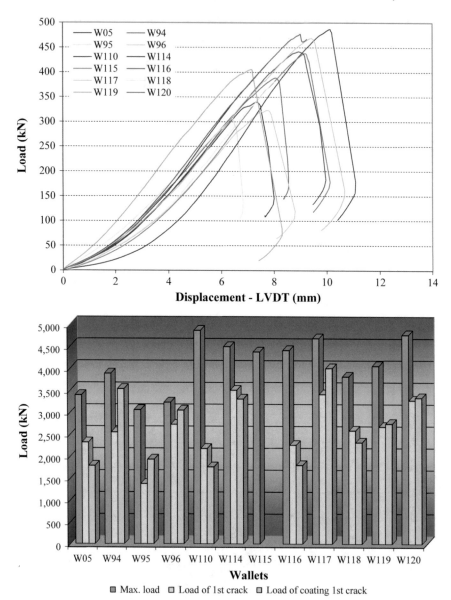

Fig. 4.29 Load versus displacement diagram of wallet specimens coated with a 3.0 cm in thickness mortar layer with a mixture proportion of 1:1:6 with reinforcement

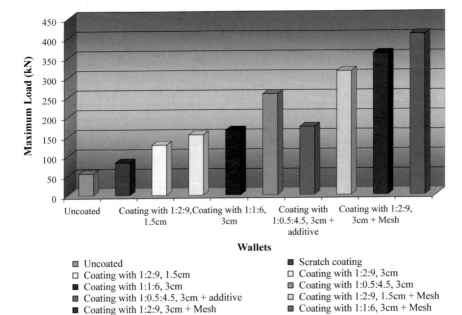

Fig. 4.30 Resume of the experimental results obtained with different wallets

References

NBR 15270-1, *Ceramic Components Part 1: Hollow Ceramic Blocks for Non-Load Bearing Masonry—Terminology and Requirements*, Rio de Janeiro, Brazil (2005)

NBR 6136, *Plain Concrete Hollow Block for Reinforced Masonry—Specification*, Rio de Janeiro, Brazil (2014)

NBR 6467, *Aggregates—Determination of Swelling in Fine Aggregates—Method of Test*, Rio de Janeiro, Brazil (2009)

NBR 7211, *Aggregate for Concrete—Specification*, Rio de Janeiro, Brazil (2009)

NBR 7215, *Portland Cement—Determination of Compressive Strength*, Rio de Janeiro, Brazi (1994)

NBR 9776, *Aggregate—Determination of Fine Aggregate Specific Gravity by Chapman Ves40 sel—Method of Test*, Rio de Janeiro, Brazil (1987)

NBR NM 248, *Aggregates—Sieve Analysis of Fine and Coarse Aggregates*, Rio de Janeiro, Brazil (2003)

NBR NM 45, *Aggregates—Determination of the Unit Weight and Air-Void Contents*, Rio de Janeiro, Brazil (2006)

NBR 7219, *Aggregates—Determination of Pulverulent Materials Content—Test Method*. Rio de Janeiro, Brazil (1982)

S.T. Andrade, *Influence of Coating Characteristics on Resistance to Compression of Masonry Walls of Ceramic Sealing Blocks*, M.Sc. Thesis in Civil Engineering, University Federal of Pernambuco, Recife, Brazil (2007)

J.M.V. Mota, *Influence of Coating Mortar on Resistance to Axial Compression in Ceramic Block Resistant Masonry Prisms*, M.Sc. Thesis in Civil Engineering, University Federal of Pernambuco, Recife, Brazil (2006)

R.A. Oliveira, F.A.N. Silva, C.W. Sobrinho, *Resistant Masonry: An Experimental and Numerical Investigation of its Compressive Behaviour*, Recife: FASA, Brazil (2011)

G.N. Araújo Neto, *Influence of Coating Mortar on Resistance to Axial Compression of Reinforced Masonry Prisms of Concrete Blocks*, M.Sc. Thesis in Civil Engineering, University Catholic of Pernambuco, Recife, Brazil (2006)

A.A.C. Azevedo, *Comparative Evaluation of the Influence of the Simple and Armed Coating on the Compressive Behaviour of Prisms and Wallets of Ceramic Sealing Blocks*, M.Sc. Thesis in Civil Engineering, University Catholic of Pernambuco, Recife, Brazil (2010)

Chapter 5
Conclusions

Throughout the research, more than 500 prisms consisting of concrete and ceramic blocks, 154 ceramic blocks wallet specimens were tested. The tests performed allow the following considerations:

- The mortar coating layer contributed to increase of the load capacity of the tested models (prisms and wallet specimens);
- The incorporation of steel meshes inside mortar coating interlocked by steel connectors generated an additional increase of the load capacity of the elements tested. The connector has played a fundamental role in increasing the load capacity of the coted prisms and wallet specimens reinforced with steel meshes, without connectors; and
- Several forms of rupture of the tested prisms and wallet specimens were observed, and, this way, it could not be chosen a failure mode that representative those observed.

Without prejudice to the foregoing considerations, it is important that the following condition be observed:

- Resistant masonry (that where it is used non-structural blocks to carry load beyond its own weight) cannot be thought, in any situation, as a building process to support loads beyond its self-weight. Consistent with this assertion, all buildings constructed using this constructive technique must be retrofitting. Risk factors determined by whatever methodology should serve merely as indicators of the order or sequence in which these building must be retrofitted. They, therefore, provide only a scale of priorities for the interventions to be carried out, which should be used with caution by the recovery process manager;
- The fact that the mortar coating layer contributes to increase the load capacity of a masonry wall serves merely to explain the reasons why the respective wall did not collapse yet. It does not authorize to certify the safety of the building that suffers from congenital failure, that is, to have been executed with non-structural blocks to carry loads;

© The Author(s), under exclusive license to Springer Nature Switzerland AG 2019
J. M. P. Q. Delgado et al., *Structural Performance of Masonry Elements*, SpringerBriefs in Applied Sciences and Technology, https://doi.org/10.1007/978-3-030-03270-8_5

– The use of the data and test results of the present research in specific solutions of recovery of buildings made with non-structural blocks is of the responsibility of the designer, and the researchers do not have any responsibility for this use.

Printed in the United States
By Bookmasters